PENGUIN BOOKS

IN SEARCH OF THE BIG BANG

Dr John Gribbin trained as an astrophysicist at the University of Cambridge before becoming a full-time science writer. He has worked for the science journal *Nature*, and the magazine *New Scientist* and has contributed articles on science topics to *The Times*, the *Guardian* and the *Independent*, and has made several acclaimed science series for BBC Radio 4. John Gribbin has received awards for his writing in both Britain and the United States and is currently a visiting Fellow in astronomy at the University of Sussex. His many books include *In Search of Schrödinger's Cat*, *Stephen Hawking: A Life in Science* (with Michael White) and *In Search of SUSY*. John Gribbin is also the author of several science fiction works including *Innervisions*.

He is married with two sons and lives in East Sussex.

ALSO BY JOHN GRIBBIN IN PENGUIN

In Search of the Edge of Time
In Search of the Double Helix
In Search of SUSY
In the Beginning
The Stuff of the Universe *(with Martin Rees)*
The Matter Myth *(with Paul Davies)*
Stephen Hawking: A Life in Science *(with Michael White)*
Richard Feynman: A Life in Science *(with Mary Gribbin)*
The Little Book of Science

John Gribbin

IN SEARCH OF
THE BIG BANG

The Life and Death of the Universe

PENGUIN BOOKS

PENGUIN BOOKS

Published by the Penguin Group
Penguin Books Ltd, 27 Wrights Lane, London w8 5TZ, England
Penguin Putnam Inc., 375 Hudson Street, New York, New York 10014, USA
Penguin Books Australia Ltd, Ringwood, Victoria, Australia
Penguin Books Canada Ltd, 10 Alcorn Avenue, Toronto, Ontario, Canada M4V 3B2
Penguin Books (NZ) Ltd, Private Bag 102902, NSMC, Auckland, New Zealand

Penguin Books Ltd, Registered Offices: Harmondsworth, Middlesex, England

First published by Heinemann, 1986
This edition published in Penguin Books 1998
10 9 8 7 6 5 4

Set in 10/12.5 pt PostScript Adobe Minion
Typeset by Rowland Phototypesetting Ltd, Bury St Edmunds, Suffolk
Printed in England by Clays Ltd, St Ives plc

I am always surprised when a young man tells me he wants to work at cosmology. I think of cosmology as something that happens to one, not something one can choose.

Sir William McCrea, FRS

Contents

Acknowledgements

The roots of this book go back a long way, to the birth of my fascination with science in the early 1950s. I cannot quite recall which author first introduced me to the mystery and wonder of the Universe, but I know that it must have been either Isaac Asimov or George Gamow, since I began reading the books of both of them so long ago that I literally cannot remember ever being without them. And it was not just science, but specifically the mystery of the origin of the Universe, that fascinated me from the outset. Thanks to Gamow and his fictitious 'Mr Tompkins' I cut my intellectual teeth on the Big Bang model of the origin of the Universe and, although later on I learned of the Steady State hypothesis, it has always been the idea of the Big Bang, the idea that there was a definite moment of creation when the Universe came into being, that held my fascination. It never occurred to me that I might make a career out of studying such deep mysteries, or writing about them. Indeed, I scarcely appreciated that being an astronomer, let alone a cosmologist, was a viable job for anyone, let alone myself, until 1966. Then, just before taking my final undergraduate examinations at Sussex University, I discovered that Bill (now Sir William) McCrea was about to establish a research centre in astronomy on the campus.

That discovery changed my life. First, it led to a swift change of direction from a planned period of postgraduate research in particle physics to a year working for an MSc in astronomy in McCrea's group. Then I moved on to Cambridge, becoming a very junior founder member of another new astronomy group, Fred (now Sir Fred) Hoyle's Institute of Theoretical Astronomy, as it then was. For reasons which I have never quite fathomed myself, I somehow became sidetracked into working on problems involving very dense stars (white dwarfs, neutron stars, pulsars and X-ray sources) for my thesis, and never did do any real research in cosmology. But while in Cambridge I met

Hoyle himself, Jayant Narlikar, Martin Rees, Geoffrey and Margaret Burbidge, Stephen Hawking, William Fowler and many other eminent astronomers who were deeply immersed in problems of literally cosmic significance. I learned from them what research at this level was really like, and I learned, too, that I could never hope to achieve anything of comparable significance myself. So I became a writer, reporting on new developments not just in cosmology and astronomy but across the sciences, keeping in touch with new developments even though I was not involved in making those new developments.

When cosmology made a great leap forward in the 1980s, it came about through a marriage with particle physics, the line of work I had abandoned so lightly in 1966. After initially struggling to cope with new developments that seemed to be appearing faster than I could write about them, I had an opportunity to catch up by attending as an observer a joint meeting organized by the European Southern Observatory and CERN, the European Centre for Nuclear Research, in Geneva in November 1983. There, participants from both sides of the fence discussed the links between particle physics and cosmology. It was that meeting, and the fact that I convinced myself that I could understand *most* of what was going on there, that convinced me that I could tackle writing this book. Following the meeting, I was able to straighten out my ideas and improve my understanding of the new idea of inflation, the key to understanding the modern version of Big Bang cosmology, in correspondence with Dimitri Nanopoulos of CERN and with two of the founders of the inflationary hypothesis, Alan Guth of MIT and Andrei Linde in Moscow.

It looks as if science has achieved, in outline at least, a complete understanding of how the Universe as we know it came into being, and how it grew from a tiny seed, via the Big Bang, into the vastness we see about us. Martin Rees, of Cambridge University, has put the importance of the new work clearly in perspective. At that meeting in Geneva, in November 1983, he commented that, when asked if the Big Bang was a good model of the Universe we live in, he used to say that 'it is the best theory we've got'. That was indeed a very cautious endorsement. But now, he said in Geneva, if asked the same question he would reply that 'the Big Bang model is more likely to be proved right than it is to be proved wrong'. Coming from Rees, one of the most cautious of modern cosmologists, who makes no claim lightly, this is a much stronger endorsement of the Big Bang, and amply sufficient justification for me to proceed in writing this book!

The fact that I can understand the physics underlying these new ideas is a tribute to the skills of teachers going back to my schooldays, and to the universities of Sussex and Cambridge; to be alive at a time when such mysteries are resolved, and to be able to understand how they have been resolved, is the greatest stroke of fortune I can imagine. Maybe new mysteries will emerge to disturb the present picture, and the completeness of our understanding of the moment of creation will prove to be an illusion. But the picture today is satisfyingly complete, and I hope I can share with you, through this book, the wonder of its completeness, and of the search which led to a successful theory of the creation, less than sixty years after the discovery that the Universe is expanding and that, therefore, there must indeed have been a moment of creation.

If I succeed at all in holding your attention, that is largely because the story is so fascinating that only the most inept of storytellers could fail to make it interesting. It is also thanks to Asimov and Gamow, who first told an earlier version of the tale to me; to Bill McCrea who, by appearing on the campus at Sussex University, showed me that cosmologists were real people and that I might work alongside them; to Fred Hoyle, who established an Institute where briefly it was possible for me to mingle with cosmologists of the first rank; and to CERN, for inviting me to attend the first ESO–CERN symposium. Once the book was underway, I received direct help from Alan Guth and Andrei Linde, from Dimitri Nanopoulos, and from Martin Rees in Cambridge and Jayant Narlikar at the Tata Institute in Bombay. Bill McCrea found time in a busy life to read parts of the book in draft and to correct some of my historical misconceptions, while Martin Rees tactfully pointed out the places where my understanding of the new ideas in cosmology was still inadequate.

Many other people, listed below in no particular order, helped by providing copies of their papers and/or giving up time to discuss their ideas with me. Thanks to: John Huchra, Tom Kibble, Roger Tayler, Carlos Frenck, Vera Rubin, Frank Tipler, John Barrow, Michael Rowan-Robinson, Stephen Hawking, Jim Peebles, David Wilkinson,

Marcus Chown, John Ellis, Tjeerd van Albada, Adrian Melott, Paul Davies and John Bahcall.

No doubt some errors remain. These are entirely my responsibility. If you spot one, let me know and I will do my best to correct it in future editions of the book. But I hope that they are few enough, and minor enough, not to mar your enjoyment of the story of the search for the ultimate cosmic truth, the origin and fate of the Universe itself.

Preface to the Second Edition

The first edition of this book appeared in 1986; it was my personal response to the wave of interest in Big Bang cosmology created by the idea of cosmological inflation. Among other things, inflation requires that the Universe contains essentially enough matter to be 'closed', so that, having been born in a superhot, superdense state (the Big Bang itself) and after expanding for many billions of years it may one day recollapse into a mirror image of the Big Bang, the Big Crunch. This in turn implies that there must be something like a hundred times more matter in the Universe than we can see, dark stuff that surrounds the bright stars and galaxies.

During the late 1980s and early 1990s, inflation theory was put on an increasingly secure footing, and the search for this dark matter intensified. When the COBE satellite found ripples in the background radiation which fills the Universe and is interpreted as the afterglow of the Big Bang itself, the pattern of those ripples (now confirmed and extended by ground-based observations) exactly matched the pattern expected from the combination of the standard Big Bang model with inflation, provided that the Universe contains enough matter to be closed. This new edition of *In Search of the Big Bang* brings the story up to date, and incorporates material from my book *The Omega Point* (now out of print) and new material to tell the whole story of the life of the Universe, from Big Bang to Big Crunch. Something had to go to make room for the new material, and I have left out most of the technical discussion of particle physics, which was tangential to the main story, and which you can find in my other books.

Although the latest cosmological ideas are not yet as complete as the basic theory of the Big Bang, which still forms the bulk of the book, the fact that they cannot explain everything does not mean that the Big Bang theory is terminally flawed. Every time that some specific detail of the developing understanding of inflation and dark matter has to be revised in the light of new observations of the Universe, somebody is sure to write an obituary for the Big Bang; but such obituaries are, like the legendary example of the premature obituary of Mark Twain, exaggerated. The Big Bang theory is very much alive and well, as I hope this book makes clear. Indeed, it now goes further than before. In 1986, I could only tell you how the Universe began; now, I can give you a fair idea of how it will end, as well. I hope you enjoy the story.

JOHN GRIBBIN
1998

the strangeness of the observation that night follows day, is not that the sky is dark, but that it should contain *any* bright stars at all. How did the Universe come to contain these short-lived (by cosmological standards) beacons in the dark?

That puzzle is brought home with full force by the light of the Sun in the daytime. This represents an imbalance in the Universe, a situation in which there is a local deviation from equilibrium. It is a fundamental feature of the world that things tend towards equilibrium. If an ice cube is placed in a cup of hot coffee, the liquid gets cooler and the ice melts as it warms up. Eventually, we are left with a cup of lukewarm liquid, all at the same temperature, in equilibrium. The Sun, born in a state which stores a large amount of energy in a small volume of material, is busily doing much the same thing, giving up its store of energy to warm the Universe (by a minute amount) and, eventually, cooling into a cinder in equilibrium with the cold of space. But 'eventually', for a star like the Sun, involves a time span of several thousand million (several billion) years; during that time, life is able to exist on our planet (and presumably on countless other planets circling countless other stars) by feeding off the flow of energy out into the void.

Because night follows day, we know that there are pockets of non-equilibrium conditions in the Universe. Life depends on the existence of those pockets. We know that the Universe is changing, because it cannot always have existed in the state we observe today and still have a dark sky. The Universe as we know it was born, and will die. And so we know, from this simple observation, that there is a direction of time, an arrow pointing the way from the cosmological past into the cosmological future.

The supreme law

All these features of the Universe are bound up with what Arthur Eddington, a great British astronomer of the 1920s and 1930s, called the supreme law of Nature. It is named the second law of thermo-dynamics, and was discovered, during the nineteenth century, not by astronomical studies of the Universe but from very practical investi-

Chapter 1

The Arrow of Time

The most important feature of our world is that night follows day. The dark night sky shows us that the Universe at large is a cold and empty place, in which are scattered a few bright, hot objects, the stars. The brightness of day shows that we live in an unusual part of the Universe, close to one of those stars, our Sun, a source of energy which streams across space to the Earth and beyond. The simple observation that night follows day reveals some of the most fundamental aspects of the nature of the Universe, and of the relationship between life and the Universe.

If the Universe had existed for an eternity, and had always contained the same number of stars and galaxies as it does today, distributed in more or less the same way throughout space, it could not possibly present the appearance that we observe. Stars pouring out their energy, in the form of light, for eternity, would have filled up the space between themselves with light, and the whole sky would blaze with the brightness of the Sun. The fact that the sky is dark at night is evidence that the Universe we live in is changing, and has not always been as it is today. Stars and galaxies have *not* existed for an eternity, but have come into existence relatively recently; there has not been time for them to fill the gaps in between with light. Astrophysicists, who study the way in which the stars produce their energy by nuclear reactions deep in their hearts, can also calculate how much light a typical star can pour out into space during its lifetime. The supply of nuclear fuel is limited, and the amount of energy a star can produce, essentially by the conversion of hydrogen into helium, is also limited. Even when all the stars in all the galaxies in the known Universe have run through their life cycles and become no more than cooling embers, space, and the night sky, will still be dark. There is not enough energy available to make enough light to brighten the night sky. The oddity, gations of the efficiency of the machines that were so important during the Industrial Revolution – steam engines.

It may seem odd that such an exalted rule of nature should be the 'second' law of anything; but the first law of thermodynamics is simply a kind of throat-clearing statement, to the effect that heat is a form of energy, that work and heat are interchangeable, but that the total amount of energy in a closed system is always conserved – for example, if our coffee cup is a perfect insulator, once the ice cube has been dropped into the hot coffee, although the ice warms and the coffee cools, the total energy inside the cup stays the same. This in itself was an important realization to the pioneers of the Industrial Revolution, but the second law goes much further.[1]

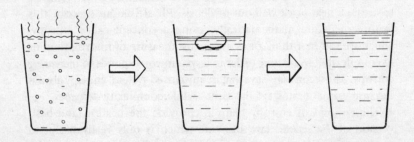

Figure 1.1 Heat always tends to even out. An ice cube placed in a beaker of hot liquid melts, and the liquid cools. We *never* see ice cubes form spontaneously out of cold liquid, while the remaining liquid heats up. This is the second law of thermodynamics, which is related to the arrow of time.

There are many different ways of stating the second law, but they have to do with the features of the Universe that I have already mentioned. A star like the Sun pours out heat into the coldness of space; an ice cube placed in hot liquid melts. We never see a cup of lukewarm coffee in which an ice cube forms spontaneously while the rest of the liquid gets hotter, even though the two states (ice cube + hot coffee) and (lukewarm coffee) contain exactly the same amount of energy. Heat *always* flows from a hotter object to a cooler one, never from the cooler to the hotter. Although the amount of energy is conserved, the distribution of energy can only change in certain ways, irreversibly. Photons (particles of light) do not emerge from the depths of space to converge on the Sun in just the right way to heat it up and drive the nuclear reactions in its core in reverse.

Stated like this, it is clear that the second law of thermodynamics also defines an arrow of time, and that this is the *same* arrow as the arrow of time defined by the observation of the dark night sky. Another definition of the second law involves the idea of information – when things change, there is a natural tendency for them to become more disordered, less structured. There is a structure in the system (ice cube + hot coffee) that is lost in the system (lukewarm coffee). In everyday terms, things wear out. Wind and weather crumble stone and reduce abandoned houses to piles of rubble; they never conspire to create a neat brick wall out of debris. Physicists can describe this feature of nature mathematically, using a concept called entropy, which we can best think of as a *negative* measure of information, or of complexity.[2] *Decreasing* order in a system corresponds to *increasing* entropy. The second law says that in any closed system, entropy always *increases* (or, at best, stays the same) while complexity *decreases*.

The concept of entropy helps to provide the neatest, and best, version of the second law, but one which is only really useful to mathematical physicists. Rudolf Clausius, a German physicist who was one of the pioneers of thermodynamics, summed up the first and second laws in 1865: the energy of the world is constant; the entropy of the world is increasing. Equally succinctly, some unknown modern wit has put it in everyday language: you can't get something for nothing; you can't even break even. This is apposite because entropy, and the second law, can also be thought of as telling us something about the availability of *useful* energy in the world. Peter Atkins, in his excellent book *The Second Law*, points out that since energy is conserved, there can hardly be an energy 'crisis' in the sense that we are using up energy. When we burn oil or coal we simply turn one (useful) form of energy into another (less useful, less concentrated) form. Along the way, we increase the *entropy* of the Universe, and diminish the *quality* of the energy. What we are really faced with is not an energy crisis, but an entropy crisis.

Life, of course, seems to be an exception to this rule of increasing entropy. Living things – a tree, a jellyfish, a human being – take simple chemical elements and compounds and rearrange them into complex structures, highly ordered. But they are only able to do so by using energy that comes, ultimately, from the Sun. The Earth, let alone an individual living thing on Earth, is *not* a closed system. The Sun is constantly pouring out high-grade energy into the void; life on Earth captures some of it (even coal and oil are stored forms of solar energy, captured by living things millions of years ago), and uses it to create complexity, returning low-grade energy to the Universe. The local decrease in entropy represented by the life of a human being, a flower, or an ant is more than compensated for by the vast increase in entropy represented by the Sun's activity in producing the energy on which that living thing depends. Taking the Solar System as a whole, entropy *is* always increasing.

The whole Universe – which must, by definition, be a closed system in this sense of the term – is in the same boat. Concentrated, 'useful' energy inside stars is being poured out and spread thin throughout space, where it can do no good. There is a struggle between gravity, which pulls stars together and provides the energy which heats them inside to the point where nuclear fusion begins, and thermodynamics, seeking to smooth out the distribution of energy in accordance with the second law. As we shall see, the story of the Universe is the story of that struggle between gravity and thermodynamics. When or if the whole Universe is at a uniform temperature, there can be no change, because there will be no net flow of heat from one place to another. Unless it contains enough matter to ensure collapse into an ultimate Big Crunch, the omega point, that will be the fate of our Universe. There will be no order left in the Universe, simply a uniform chaos in which processes like those which have produced life on Earth are impossible. Many scientists of the nineteenth century – and even later – worried about this 'heat death' of the Universe, an end implicit in the laws of thermodynamics. None seem to have appreciated fully that the corollary of the changes we see going on in the Universe is that there must have been a birth, a 'heat birth', at some finite time in the past, which created the non-equilibrium conditions we see today. And all would surely have been astonished to learn that in all but the most trivial detail the 'heat death' has already occurred.

Light and thermodynamics

Energy at high temperature is low in entropy, and can easily be made to do useful work. Energy at low temperature is high in entropy and cannot easily be made to do work. This is straightforward to understand, since energy flows from a hotter object to a cooler one, and it is easy to find a cooler object than, say, the surface of the Sun, into which energy from the Sun can be made to flow and do work along the way. It is hard to find an object colder than – say – an ice cube, so that we can extract heat from the ice cube and use it to do work. On Earth, it is much more likely that heat will flow *in* to the ice cube. Things would be a little different in space, where it is much colder than the surface of the Earth. An ice cube at 0 °C would still contain some useful energy which could be extracted and made to do work under those conditions. But still, there is a limit, an absolute zero of temperature, 0 K on the Kelvin scale named after another of the thermodynamic pioneers. An object at 0 K contains no heat energy at all.

Space itself is not quite as cold as 0 K. The energy that fills the space between the stars is in the form of electromagnetic radiation, or photons. The energy of these photons can be described in terms of temperature – sunlight contains energetic, high-temperature photons, while the heat radiated by our body is in the form of lower energy, cooler photons, and so on. One of the greatest discoveries of experimental science was made, as we shall see, in the 1960s, when radio astronomers found a weak hiss of radio noise coming uniformly from all directions in space. They called it the cosmic background radiation; the hiss recorded by our radio telescopes is produced by a sea of photons with a temperature of only 3 K, that is thought to fill the entire Universe.

This discovery, as I explain in Chapter 6, was the single key fact that persuaded cosmologists that the Big Bang theory is a good description of the Universe in which we live. Studies of distant galaxies had already shown that the Universe today is expanding, with clusters of galaxies moving further apart from one another as time passes. By imagining this process wound backwards in time, some theorists had argued that the Universe must have been born in a superdense, superhot state, the fireball of the Big Bang. But the suggestion did not meet with general acceptance until the discovery of the background radiation, which was quickly interpreted as the leftover radiation from the Big Bang fireball itself.

According to the now standard view of the birth of the Universe, during the Big Bang itself the Universe was filled with very hot photons, a sea of highly energetic radiation. As the Universe has expanded, this radiation has cooled, in exactly the same way that a gas cools when it is allowed to expand into a large empty volume (which is the basic process which keeps the inside of your refrigerator cool). When a gas is compressed, it gets hot – you can feel the process at work when you use a bicycle pump. When a gas expands, it cools. And the same rule applies if the 'gas' is actually a sea of photons.

During the fireball stage of the Big Bang, the sky *was* ablaze with light throughout the Universe, but the expansion has cooled the radiation all the way down to 3 K (the same expansion effect helps to weaken starlight, but not enough to explain the darkness of the sky if the Universe were infinitely old). The amount of everyday matter in the Universe is very small, and the volume of space between the stars and galaxies is very large. There are many, many more photons in the Universe than there are atoms, and almost all of the entropy of the Universe is in those cold photons of the background radiation. Because those photons are so cold, they have a very high entropy, and the addition of the relatively small number of photons escaping from stars today is not going to increase it by very much more. This is why it is true to say that the heat death of the Universe has already occurred, in going from the cosmic fireball of the Big Bang to the cold darkness of the night sky today; we live in a Universe that has very nearly reached maximum entropy already, and the low entropy bubble represented by the Sun is far from being typical.

The expansion of the Universe also provides us with an arrow of time – still pointing in the same direction – from the hot past to the cold future. But there is something rather odd about all this. An arrow of time, change and decay are fundamental features of the Universe at large and of everyday things that we are used to on Earth – on what physicists call a macroscopic scale. But when we look at the world of the very small, atoms and particles (what physicists call the microscopic world, although we are really talking of things far too small to be seen even with a microscope), there is no sign of a fundamental time asymmetry in the laws of physics. Those laws 'work' just as well in either direction, forwards and backwards in time. How can this be reconciled with the obvious fact that time flows, and things wear out, in the macroscopic world?

The large and the small

In real life, things wear out, and there is an arrow of time. But according to the basic laws of physics developed by Newton and his successors, nature has no inbuilt sense of time. The equations that describe the motion of the Earth in its orbit around the Sun, for example, are time symmetric. They work as well 'backwards' as they do 'forwards'. We can imagine sending a spaceship high above the Earth, out of the plane in which the planets orbit around the Sun, and making a film showing the planets going around the Sun, the moons going around the planets, and all of these bodies rotating on their own axes. If such a film were made, and were then run through a projector backwards, it would still look perfectly natural. The planets and moons would all be proceeding in the opposite direction round their orbits, and spinning in the opposite sense on their axes, but there is nothing in the laws of physics which forbids that. How can this be squared up with the idea of an arrow of time?

Perhaps it is better to pinpoint the puzzle by looking at something closer to home. Think of a tennis player, standing still and bouncing a tennis ball on the ground, repeatedly, with a racket. Once again, if we made a film of this activity and ran it backwards it wouldn't look at all odd. The act of bouncing the ball is reversible, or time symmetric. But now think of the same person lighting a bonfire. He or she might start with a neatly folded newspaper, which is spread into separate sheets which are crumpled up and piled together. Bits of wood are added to the pile, a burning match is applied, and the fire takes hold. If *that* scene were filmed and projected backwards, everyone in the audience would immediately know that something was wrong. In the

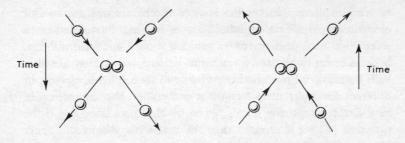

Figure 1.2 The way we imagine atoms to collide and bounce off one another in a gas, obeying Newton's laws of motion, seems to have no inbuilt arrow of time. The picture looks equally plausible whichever way we draw the arrow.

real world, we never see flames working to take smoke and gas out of the air and combine them with ash to make crumpled pieces of paper, which are then carefully smoothed by a human being and neatly folded together. The bonfire-making process is irreversible, it exhibits an asymmetry in time. So where is the difference between this and bouncing a tennis ball?

One important difference is that in the bouncing ball scenario we simply have not waited long enough to see the inevitable effects of increasing entropy. If we wait long enough, after all, the tennis player will die of old age; long before that the tennis ball will wear out (and I am not even considering the biological needs of the tennis player involving food and drink). Even the example of the planets orbiting around the Sun is not really reversible. In a very, very long time (thousands of millions of years) the orbits of the planets will change because of tidal effects. The rotation of the Earth, for example, will get slower while the Moon moves further away from its parent planet. A physicist armed with exquisitely precise measuring instruments could detect these effects from even a relatively short stretch of our film, and deduce the existence of the arrow of time. The arrow is *always* present in the macroscopic world.

But what of the microscopic world? In school, we are taught that the atoms which make up everyday things are like hard little balls, which bounce around and jostle one another in *precise* obedience to Newton's laws. Neither the laws of mechanics nor the laws of electromagnetism have an inbuilt arrow of time. Physicists like to puzzle over these phenomena by thinking about a box filled with gas, because under those conditions atoms behave most clearly like little balls bouncing off one another. When two such spheres, moving in different directions, meet each other and collide, they bounce off in new directions, at new speeds, given by Newton's laws; and if the direction of time is reversed then the reversed collision also obeys Newton's laws. This raises some curious puzzles.

One of the standard ways to demonstrate the second law of thermodynamics is with the (imaginary) aid of a box divided into two halves by a partition. Imagine one half of such a box filled with gas, and the other empty (this is only a 'thought' experiment; we don't need to actually carry it out, because our everyday experience tells us what will happen). When the partition is removed, the gas from the 'full' half of the box will spread out to fill the whole box. The system becomes less ordered, its temperature falls, and entropy increases. Once the gas is in this state, it never organizes itself so that all of the gas is in one half of the box once again, so that we could put the partition back and restore the original situation. That would involve decreasing entropy. On the macroscopic scale, we know that it is futile to stand by the box, partition in hand, and wait for an opportunity to trap all of the gas in one end.

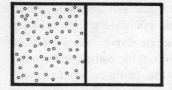 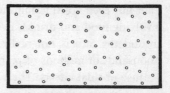

Figure 1.3 But when we look at the behaviour of a large number of atoms in a box of gas, it is easy to see the asymmetry of time. When a partition in the box is removed, the gas spreads out to fill the whole box. We can easily tell which is 'before' and which 'after', with no need for labels or arrows on the diagram. The explanation is that Newton's laws do not tell the whole story about collisions between atoms.

But now look at things on the microscopic scale. The paths followed by all of the atoms of gas in moving out from one half of the box – their trajectories – all obey Newton's laws, and all of the collisions the atoms are involved in along the way are, in principle, reversible. We can imagine waving a magic wand, after the box has filled uniformly with gas, and reversing the motion of every individual atom. Surely, then they would all retrace their trajectories back from whence they came, retreating into one half of the container? How is it that a combination of perfectly reversible events on the microscopic scale has conspired to give an appearance of irreversibility on the macroscopic scale?

There is another way of looking at this. In the nineteenth century, the French physicist Henri Poincaré showed that such an 'ideal' gas, trapped in a box, from the walls of which the atoms bounce with no loss of energy, must eventually pass through every possible state that is consistent with the law of conservation of energy, the first law of thermodynamics. *Every* arrangement of atoms in the box must happen, sooner or later. If we wait long enough, the atoms moving about at random inside the box *must* all end up in one end, or indeed in any other allowed state. Putting it another way, if we wait long enough the whole system must return once again to any starting point.

'Long enough', however, is the key term here. A small box of gas might contain 10^{22} atoms (that is, a 1 followed by 22 zeroes), and the time it would take for them to return to any initial state would probably be many, many times the age of the Universe. Typical 'Poincaré cycle times', as they are known today, have more zeroes in the numbers than there are stars in all the known galaxies of the Universe put together, numbers so big that it doesn't really make any difference whether you are counting in seconds, or hours, or years. Putting it another way, those huge numbers represent the odds against any particular state occurring, by chance, during any particular second, or hour, or year that you happen to be watching the box of gas.

This provides the standard 'answer' to the puzzle of how a world that is reversible on the microscopic scale can be irreversible on the macroscopic scale. The irreversibility, the traditionalists allege, is an illusion. The law of increasing entropy is a *statistical* law, they say, in the sense that a decrease in entropy is not so much specifically forbidden as extremely unlikely. If we watch a cup of lukewarm coffee for long enough, according to this interpretation, it will indeed spontaneously produce an ice cube while the surrounding liquid gets hotter. It just happens that the time required for this to occur is so much longer than the age of the Universe that we can, for all practical purposes, ignore the possibility.

This interpretation of the law of increasing entropy as a statistical rule, not an absolute law of nature, has recently been questioned. But long before that challenge was raised, the probabilistic interpretation led to one of the strangest theories about the origin of the Universe as we know it, and of the arrow of time – a theory well worth describing for its curiosity value, even though it is no longer taken seriously.

An improbable Universe

The idea that all states allowed by the first law of thermodynamics are constantly recurring, if we wait long enough, is hard to reconcile with the implication of the second law of thermodynamics, that entropy is increasing and that there is a unique direction in which time's arrow points. Ludwig Boltzmann, who was born in Vienna in 1844 and became one of the great developers of the ideas of thermodynamics, found a way to reconcile the two ideas. But it meant abandoning the 'common-sense' understanding of the flow of time, and also introducing the idea of a universe unimaginably more vast than anything we can see.[3]

Poincaré's work had shown that any closed, dynamic system must repeat itself indefinitely, given enough time, passing through every possible new state. This does not solely apply to boxes of gas, but to *any* system, including the Universe itself, or our Milky Way Galaxy. In a truly infinite universe, extending forever in space and with an eternal lifetime, anything which is not explicitly forbidden by the laws of physics must happen somewhere, at some time (or, indeed, in an infinite number of places, and an infinite number of times). Boltzmann's argument was that the entire observable Universe must represent a small, local region of a much bigger universe, a region in which one of those very rare, but inevitable, fluctuations, equivalent to all the atoms in the box of gas gathering in one end, or the ice cube forming in a cup of coffee, had happened, on a grand scale.

In Boltzmann's day, 'the Universe' meant our Milky Way Galaxy. It wasn't until the twentieth century that astronomers fully appreciated that our Galaxy, containing hundreds of billions of stars, is just one among many billions of galaxies scattered across a vast sea of space. But that doesn't affect the argument – it simply adds a few more zeroes to numbers already far too large for human comprehension.

The argument goes like this. Suppose that there is a universe out there which is vastly bigger than anything we can see, but which is, overall, in thermal equilibrium with maximum entropy. According to Boltzmann:

In such a universe, which is in thermal equilibrium and therefore dead, relatively small regions of the size of our galaxy [Universe] will be found here and there; regions (which we may call 'worlds') which deviate significantly from thermal equilibrium for relatively short stretches of those 'aeons' of time.[4]

The only change we have to make to bring the description up to date is the insertion of the word 'Universe'. What Boltzmann said is simply that we are living in a bubble of space where there has been a

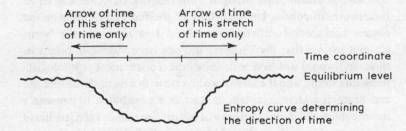

Figure 1.4 When the arrow of time is interpreted as an indicator of the direction in which entropy increases, it is possible to imagine that the Universe as we know it has been produced by a random fluctuation of entropy. In that case, wherever an observer may be in the region of low entropy, the local 'arrow of time' will point in the direction of increasing entropy. Perhaps the arrow of time is not a universal absolute, after all?

small, local deviation from equilibrium, and which is now returning to the long-term natural state of the greater universe. As Boltzmann pointed out, the arrow of time in such a low-entropy bubble will point from the less probable state to the more probable state, in the direction of increasing entropy. There is no unique arrow of time referring to the whole universe, but only a local arrow of time which applies to the region we happen to be living in. The bizarre nature of this interpretation of time's arrow (of course, Boltzmann didn't use this term, which hadn't been invented then) can best be seen from a diagram (Figure 1.4), which makes it clear that *everywhere* in the bubble of low entropy the arrow points towards the high entropy state.

According to this point of view, the Universe is an extraordinarily improbable and unlikely event, which has inevitably happened in an infinite universe. Boltzmann's own motivation in putting the idea forward is clear from his own words:

It seems to me that this way of looking at things is the only one which allows us to understand the validity of the second law, and the heat death of each individual world, without invoking a unidirectional change of the entire universe from a definite initial state to a final state.[5]

But that is *exactly* how modern cosmologists view the Universe! Boltzmann, of course, knew nothing of the Big Bang theory or the cosmic background radiation, and lived at a time when it 'went without saying' that the Universe did not have a definite origin in time, and would not have a definite end. Today, most cosmologists think differently, and the idea of a universe with a birth, finite lifetime and death is widely accepted, at least as a possibility. Boltzmann's improbable universe is a curiosity of history, one made even less likely as a description of the real world, as we shall shortly see, by the new interpretation of thermodynamics. But his discussion of what we now call the arrow of time raises an interesting point that is well worth going into in a little detail here, and which should be borne in mind throughout the rest of this book.

There is a fundamental difference between the concept of an arrow of time, pointing in a certain direction, and our subjective impression of a *flow* of time, moving in a certain direction. The point is implicit in Boltzmann's discussion, but it has been made more forcefully in recent years by Paul Davies, of the University of Adelaide. Davies has used the example of a movie film, like our imaginary film of a fire being started. He points out (for example, in *Space and Time in the Modern Universe*) that if such a film of a time-asymmetric process were made, and then the film was cut up into its individual frames, and these were jumbled together, it would still be possible to replace them in their correct order by studying the differences between the individual pictures. It isn't necessary for the film to 'run', or for time to flow, in order for the inherent asymmetry to be apparent. The flow of time is a psychological phenomenon, which arises from the way we interact with a time-asymmetric Universe.

The analogy Davies uses to help make this clear is with a compass needle on board a ship at sea. The needle always points to the north magnetic pole, indicating an asymmetry. But that does not mean that the ship is always sailing north. The ship could be sailing due south (or in any other direction) and the arrow would still point north. Or, if we wished, we could choose to make our compass needles so that the arrows on them point south. They would still be just as useful for navigation, even though the arbitrary convention assigning a 'direction' to them had been reversed. It is natural that we should define the direction in which the arrow of time points as the direction in which we perceive time to flow. But it is important to remember that the asymmetry is a built-in feature of the Universe, while our perception of the flow is a phenomenon that nobody can claim to understand. In particular, if time 'flowed' backwards it would make no difference to the asymmetry, and all of the thermodynamic arguments would still stand.

It may seem like philosophical hair-splitting, but hold on to the idea; it's going to be important later on. First, though, let's look at those new ideas in thermodynamics, which turn the traditional interpretation on its head, pull the rug from under Poincaré and Boltzmann, and hold that the irreversibility really is a fundamental feature of our Universe, once the arrow of time is defined.

The irreversible Universe

The new ideas stem mainly from the work of Ilya Prigogine, who was born in Moscow in 1917 but has been associated with the Free University of Belgium since 1947, and more recently also with the University of Texas, Austin. He received the Nobel Prize for Chemistry in 1977, for his work on non-equilibrium thermodynamics. But his ideas have yet to filter through into the textbooks used by most students of thermodynamics, even at university level.

Prigogine's attack on the problem of reconciling macroscopic irreversibility and microscopic reversibility can be understood in terms of the Poincaré recurrence time, by taking on board some basic ideas from quantum theory. Quantum physics, developed in the first half of the twentieth century, provides a better description of the behaviour of atoms and smaller particles than the older, classical ideas of electromagnetism and Newtonian mechanics. It is only with the aid of quantum physics that modern scientists are able to understand the workings of atoms, and the interactions between particles and electromagnetic fields. We don't need to go into all the details here,[6] but there are two key features of quantum physics that are relevant to the thermodynamics of the Universe.

The first important point is that the equations of quantum physics, like those of classical physics, are time-symmetric. There is no arrow of time built in to quantum physics, and reactions, or interactions, can, according to the equations, proceed just as happily 'backwards' as 'forwards'. That seems to leave us in the same bind that Boltzmann was stuck in with the conflict between the reversibility of Newtonian mechanics and the wearing out of the real world. But the second salient feature of the new physics gets us off that particular hook.

Werner Heisenberg, a German scientist who made major contributions to the development of quantum theory, discovered that the equations do not allow us to make a precise measurement of both the position and the momentum of a particle at the same time. We cannot know, as a matter of principle, exactly where a particle is *and* where it is going. We can determine either property *on its own* as accurately as we like, but the more precisely we measure position the less information we have about momentum, and vice versa. The same rule, incidentally, applies to other pairs of what are called 'conjugate variables', but that need not concern us for now.

When Heisenberg first reported this uncertainty principle, many people thought that he was talking about some limitation on the practical skills of human observers, and meant that although an electron, say, could be in a definite position and be moving with a definite velocity and momentum, it was forever beyond our skill to measure them both at the same time. Indeed, many people today still think that this is what quantum uncertainty is all about. But they are wrong. The essential feature of Heisenberg's discovery – indeed, in many ways the essential feature of quantum physics – is that the entity we call 'an electron' *does not possess* both a well-defined position *and* a well-defined momentum, simultaneously. There is an *intrinsic* uncertainty, which has to do with the way the Universe is put together, and has nothing to do with the skill, or otherwise, of human experimental physicists.

This is not common sense, but why should it be? Our common sense is based on everyday experience with objects on the human scale, and on that scale the uncertainty effect is far too small to notice. We have no basis to know what the 'common sense' of things on the scale of atoms and electrons is, except with the aid of theories that predict how collections of such particles will respond in certain circumstances. The theory that makes the best, most accurate and most consistently correct predictions is quantum theory, including the uncertainty principle. Indeed, this is only the tip of the iceberg of quantum oddity, for the best interpretation of the 'meaning' of quantum theory is that there is *no* underlying 'reality' which builds up to make the macroscopic world. The only reality lies in the actual events we observe – the swing of a needle across a dial when an electric current flows, the click of a Geiger counter as a charged particle passes through its detector, and so on. Nothing is real unless it is observed, say the quantum physicists, and there is no point in trying to imagine what atoms and electrons are 'doing' when they are not being monitored.

All of these ideas carry over into Prigogine's version of thermodynamics. The reality that we observe is the macroscopic world, with its inbuilt arrow of time and asymmetry. Why, he asks, should we imagine that this world is built up in some way from the behaviour of countless tiny particles obeying precisely reversible, time-symmetric laws of behaviour? To Prigogine, the macroscopically derived second law of thermodynamics is the fundamental truth, a *precise* law that always holds, not a statistical rule of thumb that applies, more or less, for most of the time. It is the apparently time-symmetric behaviour of little spheres bouncing off one another that he regards as an approximation to reality. 'Irreversibility,' he says, 'is either true on *all* levels or none. It cannot emerge as if by a miracle, by going from one level to another.' (*Order out of Chaos*, page 285.)

We can see what he is driving at, and the direct relevance of quantum physics to thermodynamics, by looking at another example of a closed system, the kind which Poincaré said ought to return to its initial conditions, given enough time. We start, once again, with a box full of gas, but this time make it a little bit more complicated, by placing in the box a smoothly sloping hill of material, completely symmetrical, rising up to a rounded top. Imagine a perfectly round ball balanced precisely on top of that hill, with the box shut, as usual, to keep it thermodynamically isolated from the rest of the Universe. What will happen to the ball? Obviously, it will roll off the top of the hill. But which way will it roll? The direction taken by the ball, and the subsequent history of the material in the box, will depend on some tiny accumulation of little nudges by the atoms of gas bouncing off it. There will be a minuscule pressure pushing the ball one way, just by chance, and off it will roll.

According to Poincaré, eventually the ball will return to its starting place. When the ball rolls off the hill, it gives up energy to the gas, energy derived from the fall of the ball, ultimately from gravity. If we wait long enough (many, many times the age of the Universe!) it will just happen that most of the atoms of gas bouncing off the ball will be moving in the same direction, and will give it a little push, with precisely the same amount of energy that it previously gave up to the gas, so that it rolls back up the hill while the gas cools down. There will be other occasions when the ball receives a push in the wrong direction, or one that is too strong, or too weak, to leave it balanced once again on top of the hill. But after a suitably long interval, there will be an occasion when the push is exactly sufficient to return the ball to the top of the hill, and leave it balanced there. The system has returned, as predicted, to its original state. Or has it?

If there is the tiniest difference in the way the atoms of gas now strike the ball, compared with the first time it was on top of the hill, it will roll off in a different direction, and the future history of the little world inside the box will be completely different. And there *must* be tiny differences in the way the atoms are striking the ball, because quantum uncertainty makes it impossible to define *any* set of conditions precisely for the atoms. Even in this very simple case, we can imagine the ball so precisely balanced that as tiny a change in the conditions as you like will alter its future behaviour. The real Universe is vastly more complicated than this, and complex systems involving many particles are known to be prone to very strong instabilities, so that a tiny change in starting conditions produces a drastic alteration in the system's future behaviour.

Or, if you prefer, think of things in terms of the actual reversibility of atoms moving in a box of gas. When we think about the system where the gas in half of the box spreads out to fill the whole box, it is easy to say 'imagine reversing the motion of every atom simultaneously'. The image this conjures up is of something like a pool table, with balls moving on it, which suddenly reverse their trajectories and return to their starting positions. We can't actually do the trick, but we can, indeed, imagine it. But think what the simple statement really means. It requires that the position of *every* atom should be precisely determined, and that its velocity should *simultaneously* be precisely determined, and then exactly reversed while the atom stays in precisely the same position. But quantum physics tells us that this is impossible! No atom does have the two characteristics (precise position and precise velocity) at the same time! The laws of nature, as they are best understood today, make it impossible to reverse the direction of every atom in the gas, *in principle*, not just because of the practical limitations set by human skills.[7] There is no magic powerful enough to do the job. So, once again, we find that a system that seems to be reversible is, in fact, irreversible.

This is just a very simple interpretation of one aspect of Prigogine's re-interpretation of thermodynamics. But the gist of his message is plain, and important. *No matter how long* we sit and watch a lukewarm cup of coffee, it will *never* spontaneously give birth to an ice cube and heat up; no matter how long we sit by a box of gas, it will *never* all congregate in one half of the box, so that we can trap it in a state of lower entropy. The second law of thermodynamics is an absolute ruler of the Universe.

These are difficult ideas, made no easier to absorb by the tie-in with the quantum theory idea that there is no underlying reality to nature. To Prigogine, reality lies only in the irreversible processes going on in the world – not in the 'being', as he puts it, but in the 'becoming'. Is Prigogine's version of thermodynamics better than the traditional, standard version? As of now, it is largely a matter of personal prejudice which version you prefer; but Prigogine has one powerful argument in his favour, sufficient to persuade me to side with him until he is proved wrong – nobody has ever seen a violation of the second law of thermodynamics, and until they do it seems best to accept it as just that, an unbreakable *law* of nature.

For the story of the Universe at large, though, that is as far as we need take the debate for now. We *do* live in a region of increasing entropy, an expanding bubble of dark space dotted with a few bright lights, the stars and galaxies. Time *does* flow, as we perceive it, forwards from the Big Bang to the omega point, the death of the Universe. The ultimate fate of the Universe, it turns out, depends on just how much matter it contains, not only in those bright stars but in dark forms between the stars and galaxies. But it also depends on how the Universe got to be the way it is today – and the present 'best buy' among the cosmological theories is a far cry indeed from Boltzmann's vision of an enormous Poincaré cycle. To know the future we must understand the past. To understand the nature of the Universe we live in, and to gain some insight into its likely fate, we must understand, as best we can, where it came from, and how it came to be as we see it today. And that understanding begins with observations of the Universe at large – the realm of the nebulae.

Chapter 2

The Realm of the Nebulae

The Universe in which we live is a huge and almost empty place. Bright stars, like our Sun, huddle together in groups called galaxies, which may contain a trillion (million million) stars. Some idea of how distant the stars are from each other can be gained by looking up at the dark night sky, and realizing that each of the tiny pinpricks of light we see represents a star, and that each of these stars is intrinsically about as bright as our Sun. When we look up at the night sky on the darkest, moonless night, far away from the city lights, we can see no more than two thousand stars with the naked eye. The faint band of light which we call the Milky Way is all that marks the rest of our island in space, the combined glow of millions upon millions of stars too faint and far away to be seen separately except with the aid of a telescope.

And yet, that Milky Way Galaxy of stars is itself no more than an island in space, a dot on the cosmic landscape. Just as there are millions upon millions of stars in our Milky Way Galaxy, so there are many millions of other islands in space, other galaxies, scattered across the Universe, separated from each other by distances hundreds or thousands of times greater than the size of the whole Milky Way.

To a cosmologist, someone who studies the nature and evolution of the whole Universe, a galaxy is just about the smallest thing worthy of consideration. To a human being, living on one small planet circling an ordinary star in the backwoods of a run-of-the-mill galaxy, a galaxy – our Milky Way – is just about the largest thing we can have knowledge of with our own senses, and then only with one of them, sight.

The best modern scientific understanding of the Universe reveals that it was born in fire, a Big Bang of creation some fifteen billion years ago. Cosmologists can now explain how the Universe got a superhot, superdense fireball into the state we see today, with island galaxies separated from one another by vast gulfs of space. They can, thanks to the very latest work by researchers such as Stephen Hawking, in Cambridge, at least suggest how and why the Big Bang itself occurred. And they can provide us with at least an outline guide to the ultimate fate of the Universe. This understanding of the origin and fate of the entire Universe, of everything that exists and of which we can have knowledge, is the theme of my book. But none of this understanding could have been achieved without the discovery that our Milky Way is just one, ordinary galaxy amongst millions. The scientific search for an understanding of the origin of our Universe – the search for the Big Bang – really began when other galaxies, beyond our Milky Way, were first unequivocally identified as comparable collections of stars to our own Galaxy. And that firm identification was made only in the 1920s.

Cosmology is very much a science of the twentieth century. But like all of twentieth-century science, its roots go back to the speculations of natural philosophers and metaphysicians of old.

An original theory

To the ancient Greeks and Romans, the Earth was both the centre of the Universe and its most important constituent. Although Greek philosophers had a fair grasp of the distance to the Moon, it was only with the advent of the telescope in the seventeenth century that anyone began to comprehend the great remoteness of the stars. Galileo was the first person to use a telescope for astronomical observations, and he was surprised to discover that even with the aid of his telescope's magnifying power the stars still appeared only as points of light, not as spheres like the Sun and planets. This could only mean that they were very much further away than the Sun and planets. He also found a multitude of stars visible through the telescope but unseen to the unaided human eye, and his telescope revealed the Milky Way itself to be made up of swarms of individual stars. At the same time, in the early seventeenth century, that Galileo was opening a new observational window on the Universe, Johannes Kepler was developing the basis of a theoretical understanding of our own backyard, the Solar

System. His discovery of a relationship between the time it takes for a planet to orbit once around the Sun and the average distance of that planet from the Sun led, by the 1670s, to a reasonably accurate estimate of the distance from the Earth to the Sun, which we now know to be about 150 million kilometres. Kepler's observations also provided one of the foundations for Isaac Newton's study of gravity.

It took another 150 years for astronomers to refine and improve both their observations and their theories to the point where accurate distances to a few stars were first estimated, in the late 1830s. Such estimates, and those of the twentieth century, provide a crucial stepping stone in measuring the scale of the Universe, right out to the most distant galaxies. But even before the distances to the stars were known accurately, the revolutionary discoveries of the seventeenth century provided a new view of the Universe, on a vastly greater scale than the old vision of a series of crystal spheres surrounding the Earth and extending out a little beyond the orbit of Saturn. In the eighteenth century, a few philosophers interpreted these new discoveries in terms of a picture, an imaginary model, of the Milky Way and its place in the Universe. That model is surprisingly close to modern thinking, and fuelled debate among astronomers and philosophers for the best part of two centuries.

Credit for the new theory of the Universe – the first modern cosmological theory – belongs to Thomas Wright of Durham, England. Wright was an English philosopher, born in 1711, who like most thinkers of his era spread his interests over a wide variety of subjects, including astronomy. He was the son of a carpenter, and his interest in astronomy was kindled by his childhood teacher. But his formal education was curtailed by a serious speech impediment and for a time he ran wild, becoming, he tells us in his journal, 'much addicted to sport'. At thirteen he was apprenticed to a clock- and watchmaker, where he stayed for four years but spent all of his spare time studying astronomy, encouraged by his mother but violently opposed by his father, who did everything he could to prevent these studies, including burning young Thomas's books. During his turbulent early adult years, Wright tried his hand at the sailor's life, quitting after a violent storm during his first voyage, set up as a tutor of mathematics at Sunderland, was involved with a scandal concerning a clergyman's daughter, taught navigation to seamen, and then in the 1730s began to achieve success and prosperity (after some initial trials and tribulations with dishonest publishers and a failed attempt to produce an almanac) as a tutor and consultant to the aristocracy. The speech impediment, if it still existed, was no longer a handicap to this confident and self-assured young man. He would survey a grand estate (or a modest one), hold private classes in natural philosophy, mathematics or navigation, and along the way began to publish successful books and broadsheets. By 1742, his reputation was such that Wright was invited to become Professor of Navigation at the Imperial Academy in St Petersburg, at a salary of £300 a year; he declined the post after failing to get the proposed salary increased to £500 per annum. So it was as a successful, educated (if largely self-educated) and reasonably well-known philosopher of his day that Thomas Wright published, in 1750, a work entitled *An Original Theory or New Hypothesis of the Universe*. This is the work for which he is remembered today, with a place in the history of science important enough for the book to have been reprinted, in facsimile form, in 1971.

In the early seventeenth century, a hundred years after Galileo's revolutionary discoveries, it was widely accepted among natural philosophers that the stars must be distant bodies that shine with their own light, like the Sun, and not because (like the Moon) they reflect the light from the Sun. Since telescopes could not show the stars as discs, they must be very far away indeed. Many, but not all, of Wright's contemporaries would argue that since the brightest object we know is the Sun, therefore the best guess we can make about the stars is that they are comparably bright objects set at suitably remote distances from us. Some speculated that the apparent randomness of the pattern of stars on the sky was simply due to us being embedded within a system whose pattern and structure could only be seen from outside – like the shape of a forest which is concealed from someone inside the forest, who sees only a seemingly random distribution of trees in all directions.

In earlier works, described by Michael Hoskin in his introduction to the 1971 edition of *An Original Theory*, Wright considered the Universe (as we would now call it) like a sphere or globe filled with stars, and described how an observer sitting on a planet like the Earth orbiting one of those stars would see nearby stars bright and clear, others further away too faint to be resolved without the aid of a telescope, and still others, yet more distant, so faint that they appear only as a band of light circling the sky. The description begins to fit the Milky Way, except, as Wright realized later if not sooner, that if we are surrounded on all sides by an even distribution of stars then the whole sky should appear to glow like the band of the Milky Way. By 1750, in *An Original Theory*, Wright had hit on the reason why the Milky Way appears as a band across the sky. We must imagine the stars, he says, 'all moving the same Way, and not much deviating from the same Plane, as the Planets in their heliocentric Motion do around the solar Body.' In other words, our Sun is but one star in a great swarm which fills a flat disc, not a spherical volume of space. When we look along the disc, towards the centre around which all the stars orbit, we see the profusion of stars that together form the band of light we call the Milky Way. But when we look out of the thin disc upwards or downwards into the depths of space, we see only a few nearby stars and no band of light because there are no very distant stars to see. In fact, Wright's vision of the star system was more like the rings of Saturn, with a central gap, than like a solid disc of stars. And, to be honest, his speculations about the nature of the Milky Way did not form what he regarded as the most important of his contributions to science and philosophy, since his main interests concentrated on topics which would today be regarded as primarily the province of religion. Indeed, what seems with hindsight to be Wright's deepest insight of all appears almost as an aside, tossed away in a summing up at the end of his book. Having offered an explanation for the appearance of the Milky Way, he went on to speculate on what might lie outside this island in space. What we now call the Galaxy Wright called the Universe, or the Creation, and he imagined all the stars with their own families of planets, sidereal systems like our own Solar System. 'As the visible Creation is supposed to be full of sidereal Systems and planetary Worlds,' he wrote, 'so on, in like manner, the endless Immensity is an unlimited Plenum of Creations not unlike the known Universe.' In other words, Wright imagined the immensity of space beyond our Milky Way Galaxy to be populated by other galaxies like the Milky Way. And he even suggested that faint patches of light on the sky, called nebulae, visible with the aid of telescopes but unresolved into stars, 'may be external Creation[s] [galaxies], bordering upon the known one, too remote for even our Telescopes to reach.'

With these words, Wright lit a spark of understanding concerning the nature of our Galaxy and of the Universe, a spark which was gradually fanned into a flame, initially through the work of another philosopher, Immanuel Kant, and another astronomer, William Herschel.

A nebular hypothesis

Kant is best known as a philosopher, one of the first rank, whose teaching has had an influence on all philosophical thinking of the past two hundred years. But even so, it is surprising to find his entry in the *Encyclopaedia Britannica* scarcely mentioning his interest in astronomy. Even more surprisingly, there is no mention at all of Kant's early work *General History of Nature and Theory of the Heavens*, which was published in 1755. Greatly influenced by Wright's *An Original Theory*, which had appeared five years before, in his own *Theory of the Heavens* Kant presented speculations about the nature of the Universe which are both much clearer than Wright's ideas and bear an almost prescient resemblance to the present picture of the Universe, which is based on twentieth-century observations. Kant never performed experiments or observations, and from time to time over the years has been derided as an 'armchair scientist' who simply speculated about the meaning of discoveries made by others. If only there were a few more armchair scientists like him – and like Einstein, who also relied solely on brain power, not on any practical skills.

Kant was born in what was then Königsberg, East Prussia (now Kaliningrad, in the Russian republic), in 1724. His surname came from a Scottish immigrant called Cant, his grandfather, who settled there and Germanicized his name; Immanuel Kant's father was a saddle-maker, and the boy himself was the fourth of eleven children in a poor family, the oldest child to survive to maturity. His parents were devout Lutherans, and it was through the influence of their pastor that young Immanuel took the first steps towards a formal education. Indeed, in 1740 he enrolled at the University of Königsberg as a theological student; but it was courses in mathematics and physics that caught his fascination and to which he devoted most of his attention. He was by then set upon an academic career, but when his father died in 1746 Immanuel had to leave the university and work as a private tutor – by no means uncongenial employment, since his three employers over the span of the next nine years were all influential society families who introduced him to a new way of life and circle of acquaintances. It was during this period that Kant made the longest journey of his life, travelling sixty miles to the town of Arnsdorf.

In 1755, Kant was at last able to take up a post at the university, as a *Privatdozent*, or lecturer. In spite of offers of good posts at other universities, he remained in Königsberg all his life, becoming Professor of Logic and Metaphysics in 1770 and holding the Chair for twenty-seven years. He died in 1804. But the work for which he will always be remembered by cosmologists was completed while he was merely a private tutor, and was published in the year that he first took up an official university post.

Wright's *An Original Theory* never enjoyed wide circulation, and soon became quite rare. Kant himself, as far as we know, never saw the book itself, but learned of Wright's ideas through a lengthy review of the book, which appeared in a Hamburg journal in 1751. Happily, the review Kant read, which included quotes from Wright, was not only accurate but in some ways clearer than Wright's own book, since it contained only the central kernel of Wright's ideas, including the idea of the Milky Way as a collection of stars 'all moving in the same way and not much deviating from the same Plane'. Kant elaborated the theme in his own book, and made much stronger statements than Wright about the likely nature of the faint patches of light, the nebulae, that could be seen with telescopes but clearly were not individual stars. In his book *The Realm of the Nebulae*, Edwin Hubble, one of the twentieth-century pioneers of cosmology, provides a good translation of a key passage from *Theory of the Heavens*:

I come now to another part of my system, and because it suggests a lofty idea of the plan of creation, it appears to me as the most seductive. The sequence of ideas that led us to it is very simple and natural. They are as follows: let us imagine a system of stars gathered together in a common plane, like those of the Milky Way, but situated so far away from us that even with the telescope we cannot distinguish the stars composing it ... such a stellar world will appear to the observer, who contemplates it at so enormous a distance, only as a little spot feebly illumined and subtending a very small angle; its shape will be circular, if its plane is perpendicular to the line of sight, elliptical, if it is seen obliquely. The faintness of the light, its form, and its appreciable diameter will obviously distinguish such a phenomenon from the isolated stars around it.

We do not need to seek far in the observations of astronomers to meet with such phenomena. They have been seen by various astronomers, who have wondered at their strange appearance.[1]

Hubble was particularly pleased with Kant's reasoning because it is based on what is now known as the principle of uniformity, or, slightly more tongue in cheek, the principle of terrestrial mediocrity. This says that we live in a typical, ordinary part of the Universe, and that any other typical, ordinary part of the Universe would look much the same as our neck of the woods. We live in an *ordinary* galaxy, says Kant by implication, and the faint nebulae seen by the astronomers are no new phenomenon but other *ordinary* galaxies like our own. Anyone living on a planet circling a star in one of those galaxies would see our Milky Way as no more than a faint patch of light barely visible through their telescopes. And as if credit for presenting the idea itself clearly for the first time were not enough, Kant is also generally credited with inventing the term 'island universes' to describe the way galaxies are scattered across the immense void of space.

Kant, and Wright (whose work remained for two hundred years best known through the references in Kant's writing) got the kudos, but this was clearly an idea whose time had come. Johann Lambert, a Swiss-German polymath who among other things provided the first rigorous mathematical proof that *pi*, the ratio of the circumference of a circle to its diameter, is an irrational number, independently came up with a similar scheme of the Universe, and published it in 1761. Indeed, according to a letter he wrote to Kant (and we have no reason to doubt his word) Lambert first began to speculate along these lines in 1749. It was only after his own paper had been prepared that he learned of the work of Wright and Kant. From the 1760s onward, however, the island universe idea became commonly known among astronomers, although not yet generally accepted. The armchair scientists had taken their thoughts as far as they could reasonably go with the limited observations of the heavens available to them, and it was time for the observers, once again, to come to the centre of the astronomical stage. Not until the twentieth century would the observations be adequate to provide the final test of the hypothesis that nebulae are galaxies in their own right, and then to provide, in an increasing flood, new waves of information for the theorists in their armchairs to digest and try to make sense of.

Observing the Universe

Few eighteenth-century observers were much interested in the nebulae in their own right, but many astronomers were interested in comets. The discovery of a comet, apart from its scientific importance, brought instant fame to any astronomer, and a 'new' comet was (and still is) traditionally named after its discoverer. Edmond Halley, whose name is forever linked with comets, had died only in 1742, leaving untested his claim that cometary sightings in 1456, 1531, 1607 and 1682 all represented visits of the same comet to the inner part of the Solar System, and his prediction that the comet would return again in 1758. When it duly appeared on schedule, Halley's place in history was assured, as was the eagerness with which astronomers of the second half of the eighteenth century would seek out other comets. But one of the irritating things about comets is that when they are first identifiable they appear as faint little patches of light in the field of view of a telescope, just like the nebulae which Kant found so fascinating. Many an astronomer thought he had hit the jackpot, only to have his hopes of fame dashed when observations on succeeding nights showed that the patch of light he had discovered showed no signs of growing bigger and brighter as it approached the Sun, but stayed always the same. What astronomers needed was a catalogue of all the known, permanent nebulae so that they wouldn't be fooled into mistaking them for new comets. And the first decent catalogue of this kind was provided by Charles Messier, a French astronomer, who made his painstaking compilations between 1760 and 1784. The brightest nebulae, together with clusters of stars, were pinpointed and identified not because Messier regarded them as particularly important, but because they were an annoyance that had to be signposted 'keep off, this is not a comet'. The catalogue served its purpose well; Messier himself discovered at least fifteen comets (some reports claim as many as twenty-one). And when an astronomer came along who *was* interested in nebulae in their own right, Messier's catalogue gave him an invaluable starting point for his own observations. That astronomer was William Herschel.

Herschel was born in Hanover in 1738. Like his father, he was a musician in the band of the Hanoverian Guards, and he visited England with the band in 1756. A year later, a little trouble with the French led to the occupation of Hanover, and Herschel left to take up permanent residence in England, where he worked as a music teacher, performer and composer, until in 1766 he was appointed organist at the Octagon Chapel in Bath. Gradually, his interest in astronomy, originally no more than amateur dabblings, became a passion, and he determined not just to make observations of the Sun, Moon and planets, like most astronomers of the time, but to study the faintest and most distant objects he could. This passion forced him to become an expert telescope maker, polishing his own large mirrors to manufacture instruments capable of showing him astronomical objects fainter than anyone had ever seen, and of showing new details in known objects. His sister Caroline, who had joined him in Bath in 1772, shared this passion for astronomy, and worked as his assistant, both in making the telescopes and in carrying out the observations. Together, they studied the skies systematically, surveying in all directions. And in one of the sweeps of their heavens, in 1781, Herschel hit the jackpot. What he thought at first must be a comet turned out to be a planet that nobody had seen before, the first to be identified since ancient times. Not one to miss an opportunity, Herschel suggested naming the new planet 'George's Star', after King George III; in the end, the name Uranus was adopted, but the King was suitably flattered and in 1782 Herschel was appointed court astronomer (not the same thing as Astronomer Royal, a post held at the time by the Reverend Nevil Maskelyne). He became a Fellow of the Royal Society in the same year, moved from Bath to Windsor and then to Slough, and became a professional astronomer, patronized by the King, at the age of forty-three.

From now on, Herschel had little difficulty raising money for bigger and better telescopes with which to probe the night skies. He had been given an early list of Messier's nebulae, and set out, with Caroline, to make his own compilation of nebulae. Their searches resulted in a catalogue listing more than two thousand nebulae in 1802, and an even more extensive catalogue published in 1820, by which time Herschel was in his eighties. He was knighted in 1816. Along the way, Herschel did much to promote the idea of island universes. He may well have got the idea direct from Wright's book, since the copy he owned still exists, and carries a note in his own hand, made some time after 1781. By about 1785, Herschel was convinced that all nebulae were composed of stars, much as Kant had speculated, and he and Caroline had been able to resolve many of these clouds into their component stars using their new telescopes.

In an attempt to explain why stars should be gathered together in this way, William developed a theory of the evolution of the Universe which involved the idea of widely scattered stars gathering slowly into clumps under the attraction of gravity. This was one of the first attempts to describe the Universe in terms of change, and although his specific idea does not stand up to modern scrutiny, Herschel deserves credit for having the imaginative boldness to conceive of the Universe itself as changing and evolving as time passes.

Unfortunately, a lot of Herschel's good work on behalf of the island universe hypothesis was undone later in his life, when he found that some of the clouds in space, some of the nebulae, could not be resolved into stars even with his best telescopes. This lent weight to the rival theory, which held that the nebulae really were just clouds of material glowing with their own light. He found that some of these clouds were associated with individual stars, a central star being surrounded by a glowing nebula, and this reinforced the idea that nebulae might be planetary systems in the making, stars and planets condensing out of a collapsing cloud of gas. Because of Herschel's enormous prestige, this idea gained great currency in the early nineteenth century; and because this theory was regarded as a rival to the concept of nebulae as island universes of stars, that theory suffered a decline in standing. It took another hundred years for astronomers to appreciate fully that there are two different kinds of nebulae, one kind glowing clouds within our own Milky Way system, and the other kind island universes, other galaxies, far beyond the Milky Way.

Bigger telescopes and new astronomical techniques began to make the true situation clear in the second half of the nineteenth century. William Parsons, the third Earl of Rosse, was an Irish politician, engineer and astronomer, born in 1800, who had an ambition to carry on observations of nebulae where Herschel had left off. Since Herschel had left no notes on how he had built his telescopes, Rosse had to reinvent most of the techniques of polishing and preparing large mirrors for himself, but his efforts culminated in the construction of a telescope weighing four tons, with a principal mirror seventy-two inches across housed in a tube more than fifty feet long, supported between two masonry piers by a system of chains and pulleys. With this magnificent instrument, dubbed 'the Leviathan of Corkstown', Rosse and his assistants had resolved fifty nebulae into stars by 1848, and noted that some of them had a characteristic spiral shape, like a whirlpool viewed from above. These discoveries strongly revived the idea of island universes. A few years later came the first hard evidence that nebulae come in two varieties.

William Huggins, a British astronomer born in 1824, pioneered the use of spectroscopy in astronomy. A spectroscope is an instrument which splits light up into its component colours, like a rainbow, spreading out the light with different wavelengths for examination. As well as the different colours of the rainbow, such a spectrum characteristically shows a pattern of bright or dark lines, very sharply defined at specific wavelengths in the spectrum. These spectral lines are caused by the presence of particular elements in the material that the light is coming from – yellow street lights, for example, produce a very bright pair of lines associated with the element sodium. No other element produces an identical pair of lines in exactly the same place in the spectrum. Spectroscopy provided a way to identify the elements present in the Sun and stars, by identifying the lines in the spectra obtained from different celestial objects. Huggins turned to this line of research in the 1860s, following the work of Gustav Kirchhoff and Robert Bunsen, who made spectroscopic studies of the Sun in the late 1850s. Together with a chemist friend, W. A. Miller, Huggins showed that starlight contained the same spectral lines as sunlight. In other words, stars contain, by and large, the same mixture of chemical elements as the Sun does.

When Huggins turned his attention to the nebulae, however, he found a different pattern. Some nebulae, like the one in the constellation Orion, or the one known, from its shape, as the Crab Nebula, glowed with a light that did not have this by now expected pattern of lines. Instead, the light from these nebulae appeared in the spectroscope just like the light from a glowing cloud of hot gas. Only later did similar studies show that other nebulae, including the spirals, did indeed produce light with spectral lines like those seen in starlight.

The pieces of the puzzle were beginning to pile up on the table, but they were not yet falling into place. Kant's ideas, now more than a century old, were, after all, no more than the musings of an armchair scientist. The gaseous nebulae were clearly part of the Milky Way, and even if the other nebulae were made of stars, they might well be part of the Milky Way too, small star systems, or stars in the process of formation, not whole galaxies as big as our Milky Way. And, anyway, how big *was* the Milky Way? At the end of the nineteenth century, the scale of the Milky Way system itself was only known in the most approximate terms, partly from estimates of how far away the stars must be in order to appear as faint as they do. The balance of opinion held that the Milky Way *was* the Universe, although it did seem to be a flattened, disc-shaped structure of the kind Kant had proposed. The Sun and Solar System were placed, in the astronomers' minds, somewhere near the centre of this disc, and the spiral and elliptical nebulae were considered, by those who bothered much about them at all, to be part of the Milky Way system. For a little over a hundred years, astronomers had indeed been observing the Universe at large, even though they could not be sure of this. The nebulae, we now know, are the building blocks of the Universe. But in order to find out their true nature, a step beyond simple observation had to be taken. The astronomers now had to measure the Universe, to get a handle on the scale of distances in the Milky Way and beyond, before they could truly grasp the significance of what it was they were observing.

Chapter 3

How Far is Up?

Edmond Halley was the first astronomer on record as realizing that the stars move. Because the stars move relative to one another, they cannot be objects with different brightnesses all attached firmly to the inner surface of some great sphere surrounding the Earth. Evidence that stars move is also evidence that stars are at different distances from us, spread out in three dimensions of space. In the early eighteenth century, when Halley made this discovery, it was the first direct, observational evidence that the image of stars as such lights attached to a sphere little bigger than the orbit of Saturn (Uranus had not then been discovered) must be incorrect. The discovery paved the way for later eighteenth-century thinkers such as Wright and Kant to make their speculations about the nature of the Milky Way; it also led, along a direct but slow path, to an understanding of the scale of the Universe.

Halley's Universe

By the time he made this discovery, Halley was already a respected senior astronomer with a fruitful career behind him. He was born in 1656, and went to Oxford University, where he wrote and published a book on the laws discovered by Johannes Kepler, which describe the orbits of the planets about the Sun. Kepler's laws gave Isaac Newton crucial clues about the nature of gravity; Halley's book was impressive enough (not least coming from an undergraduate) to bring its author to the attention of John Flamsteed, who was the Astronomer Royal (the first one, and actually called 'Astronomical Observator') at the time. This was just as well, for when Halley left Oxford without finishing his degree Flamsteed's interest helped him to get a job in astronomy, sent to the island of St Helena in the South Atlantic, to spend two years mapping the stars of the southern hemisphere sky. He returned to England in 1678, and was promptly elected to the Royal Society at the age of twenty-two. But Halley did not follow an exclusively academic career.

His adventures over the next thirty years included travels in Europe to meet other scientists and astronomers, a couple of years as Deputy Controller of the Mint at Chester, a spell in command of a Royal Navy warship, the *Paramour*, and some diplomatic missions to Vienna on behalf of His Majesty's Government. Along the way, he made major contributions to the understanding of magnetism, winds and tides, and was a key influence in persuading Newton to publish his *Principia* in its entirety – Halley even financed publication of the book. In 1703 he became Professor of Geometry at Oxford (not bad for an undergraduate dropout) and in 1720 he was appointed Astronomer Royal in succession to Flamsteed, and held the post until he died in 1742 at the age of eighty-five. It was in the two decades at the start of the eighteenth century that Halley made his contribution to our understanding of the stars.

Halley had always had an interest in the astronomical writings of the ancients, and had translated some works from the Greek. In 1710, he began a study of Ptolemy's writings, which date from the second century AD and include a catalogue of star positions. In fact, this catalogue is even older than Ptolemy, going back to the work of Hipparchus in the third century BC. This was the first important star map, containing the positions of more than eight hundred stars; Ptolemy preserved it for posterity, and added more positions to bring the number of stars logged to over a thousand. Most of the star positions in the catalogue agreed well with observations made by Halley and his contemporaries. But in 1718 Halley realized that three stars – Sirius, Procyon and Arcturus – were not in the places that Hipparchus and Ptolemy saw them. The differences in positions were much too great to be explained as mistakes by the ancient Greeks; besides, why should they make just three mistakes out of hundreds of accurate observations? Arcturus, for example, appeared in 1718 to be twice the width of the full Moon away from the position recorded in Ptolemy's writings – a full degree of arc out of position. Halley inferred that Arcturus, and the other stars, had moved over the centuries since the Greeks recorded their positions. The motion was far too slow to be noticed in a human lifetime with unaided eyes, but big enough to show up over several generations.

The first three stars to have these proper motions, as they are now called, identified are among the eight brightest stars in the sky. The natural interpretation of this 'coincidence' is that the stars look bright because they are much closer to us than most stars, and that we can see their movement over a few centuries for the same reason. Just as an airliner high in the sky appears to be crawling slowly along, while a child on a bicycle a few feet away rushes past in a flash, so the nearer stars ought to show a larger apparent motion across the sky than those further away, if they all move at more or less the same speed through space. The vast majority of stars are so far away that even over a couple of thousand years they show no apparent motion from Earth; just a few are close enough to be identified as moving across this background of seemingly 'fixed' stars.

It's probably worth pausing for a while to try to grasp the size of the movements astronomers now measure as routine. The angle the Moon covers on the sky is roughly half a degree, or 30 minutes, of arc (31 minutes to be more precise). Just as each degree is divided into 60 minutes, so each minute of arc is divided into 60 seconds. The planet Jupiter cannot be distinguished as a perceptible disc with the naked eye, but telescopic observations show that when it is at its closest to the Earth, and therefore looks biggest, it covers only 50 seconds of arc. The star which has the largest measured proper motion of all is called Barnard's Star, after the American astronomer Edward Barnard, who discovered it in 1916. Barnard's Star is faint because it really is a dim little star, which is why it wasn't noticed before. But it is also close to us – so close that it hurtles across the sky at a record-breaking speed of 10.3 seconds of arc per year. That is, in five years it moves a distance equivalent to the angular width of Jupiter at its biggest. And this is a record breaker – measured proper motions are usually less than one second of arc per year. In order to translate these tiny proper motions into speeds through space, astronomers had to find a way to estimate the distances to the stars; but first they had to get an accurate measure of the scale of the Solar System.

From the Earth to the Sun

The first step towards measuring the scale of the Universe uses exactly the technique used by mapmakers on Earth to measure positions and distances – triangulation. Astronomers usually use another name for the technique – parallax – but it is the same thing and you can see how it works simply by looking at your finger.

Hold a finger up at arm's length, and close one eye. Notice the position of your finger against the background of the wall of your room. Now close the eye that was open, and open the one that was closed. The finger seems to move across the background, jumping to a slightly different position. The reason is that your two eyes view the finger from slightly different directions. Now try the same thing with your finger held just in front of your nose. The change in position – the parallax – is much more pronounced. The closer an object is, the bigger the parallax effect, so by measuring the parallax it is possible to work out the distance to the object. If you observe a distant object from two widely separated points you can use this effect to determine its distance. The technique works as well for a mountain or for the Moon or for the planets in the Solar System – provided that, in each case, you can get a wide enough separation between the two places where you make observations from, a long enough base for your triangle. To find the distance to the Moon, for example, all that is necessary is that astronomers in two widely separated observatories should each note the position of the Moon against the background of distant stars at the same time. As long as they know the distance between the two observatories (and take account of the curvature of the Earth, of course) they can use these observations to construct an imaginary triangle with the Moon at its top. Simple geometry then tells them how tall the triangle is – how far the Moon is from the Earth.

The technique works well for the Moon, because it is so close to us, about 400,000 kilometres away. The parallax effect is quite noticeable,[1] and although the triangle involved is rather tall and thin the angles are easily measured. Things get a little trickier when astronomers try to find out the distances to the planets using the same

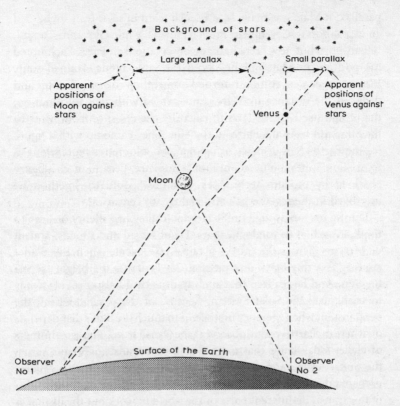

Figure 3.1 Parallax – the *apparent* movement of a distant object caused by a shift in the position the object is observed from – can be used to measure distances within the Solar System.

techniques. The parallax effect is very small, even for two observatories on opposite sides of the Earth, and the imaginary triangle drawn with the Earth at its base and Mars, say, at its apex is incredibly skinny. But still, under the right conditions the appropriate angles can be measured and the distances worked out.

Parallax measurements for the planets were impossible before the invention of the astronomical telescope, and even then they required expeditions to far-flung corners of the globe to get a wide enough base for the triangle. The first really successful measurement of the parallax of Mars was made by a French team in 1671. Jean Richer led an expedition to Cayenne in French Guiana, while Giovanni Cassini, Italian-born but now French, made observations in Paris. Each noted the position of Mars on the sky at the same predetermined time. When Richer's expedition returned home, they compared notes and deduced the distance to Mars. And, armed with *that* information, they could use Kepler's laws to calculate the distance from Mars to the Sun, and from the Earth to the Sun. They came up with a figure for the Earth–Sun distance of 140,000,000 kilometres (equivalent to 87 million miles), only 10 million kilometres less than the figure obtained by modern techniques. But how could the method be improved in the seventeenth and eighteenth centuries?

During his lonely vigil on St Helena, Halley had plenty of time to think, as well as to catalogue stars. He observed and timed a transit of Mercury across the Sun – a fairly rare event, when the planet appears as a tiny black disc passing slowly across the bright face of the Sun. And he realized that such transits could also provide a way to triangulate the Solar System. Because of the parallax effect, the exact instant when Mercury first seems to touch the Sun's disc depends on where on Earth you are observing the transit from. The astronomers of Halley's day knew that there would be a transit of Venus across the Sun – even rarer than a transit of Mercury – in 1761. Halley prepared detailed notes on just how to make best use of observations of this transit in different parts of the world to work out the distances to Venus and to the Sun, and he published the notes in 1716. Although he had been dead for nineteen years when the transit occurred, his influence was a factor in the concerted effort that took place to take the opportunity to measure the parallax of Venus. Sixty-two observing stations monitored the transit of 1761, and a similar effort was made for a second transit in 1769. When all the data had been analysed the calculated distance from the Earth to the Sun was 153 million kilometres, compared with the modern measurement of 149.6 million kilometres. In succeeding centuries, the measurement was refined down to its present value; but, as far as the broad picture is concerned, we can say that by the end of the eighteenth century astronomers knew the scale of the Solar System. They had measured the distance to the Sun. And this provided them with a great new opportunity. Once they knew the distance from the Earth to the Sun, they had a new baseline that they could use for triangulation and parallax. The measurement is so important to astronomy, indeed, that the distance from the Earth to the Sun is called the astronomical unit, or AU. With a baseline 150,000,000 kilometres long, it might be possible to triangulate the distances to the stars themselves.

From the Sun to the stars

It was three-quarters of a century after the transit of Venus in 1761 that astronomers at last managed to measure the parallax of a few nearby stars. The principle involved was simple enough. Since the radius of the Earth's orbit is 150 million kilometres, observations made six months apart, from opposite sides of the Sun, are at the ends of a baseline 300 million kilometres (2 AU) long (and it doesn't matter, from the point of view of getting a first estimate of the distances to the stars, whether the baseline is actually 290 million kilometres or 310 million kilometres; the answers we get will still be about right). It is a matter of simple geometry to calculate how far away a star would have to be in order to show a certain parallax displacement across such a baseline. In fact, astronomers choose to define a new length scale in these terms. One parallax second of arc, or parsec for short, is the distance to a star which would show a displacement of 1 second of arc from opposite ends of a baseline equal to the distance from the Earth to the Sun. In other words, over the 300-million-kilometre baseline of the Earth's orbit, a star 1 parsec (1 pc) away will show a parallax of 2 seconds of arc.

A parsec is a little more than 30,000 billion (a billion is a thousand million, 10^9) kilometres; it takes light, travelling at a speed of nearly 300 million metres a second, 3.26 years to cover a distance of one parsec. Or, to put it another way, 1 parsec is just under 206,265 times the distance from the Earth to the Sun, the astronomical unit. In fact, no star is close enough to show this much parallax displacement. And that is why it took until the 1830s for the first stellar parallaxes to be measured.

If you are trying to measure a change of less than one second of arc in the position of a star, obviously you need catalogues which give the positions of the stars to comparable accuracy. The best catalogue of the early eighteenth century, compiled by Flamsteed and published posthumously in 1725, gave positions to an accuracy of about 10 seconds of arc – a tremendous achievement at the time, but not yet good enough for this purpose. The third Astronomer Royal, Halley's successor James Bradley (born in 1693), devoted a massive effort to the problem, attempting to measure the parallax of a star called Gamma Draconis. He failed, but along the way he improved observing techniques, developed better instruments, and also improved astronomers' theoretical understanding of the nature of their observations. Bradley found that Gamma Draconis did seem to shift its position on the sky over the year, but not in the way predicted by parallax. He found that the same effect occurs for all stars, and eventually realized that the effect was due to the speed of the Earth's motion around the Sun. Light rays coming from a distant star seem to be tilted, because of the Earth's motion, in exactly the same way that rain falling straight down from the sky seems to be blowing in your face when you walk forward. The effect is different at different points in the Earth's orbit (different times of year) because our planet is moving in different directions at different times. Because the speed of light is so great, the effect is small – but still noticeable at the kind of accuracy of measurement involved in parallax studies.

Bradley called the effect 'aberration'; it produces a shift in the apparent positions of the stars of 20½ arc seconds over a year. The discovery confirmed, although it was not really news in the late eighteenth century, that light has a finite speed, and gave an estimate for that speed close to the present-day estimate; it also confirmed that the Earth moves through space. But still there were more disturbances for Bradley to take into account when measuring star positions, including a wobble of the Earth, which he named nutation, due to the Earth's slightly non-spherical shape. The fruits of all these efforts appeared in the form of a major new catalogue of about three thousand star positions of unprecedented accuracy – but this was only published, in two parts, in 1798 and 1805, more than thirty years after Bradley died.

Figure 3.2 Rain falling straight down from the sky seems to be blowing in your face when you walk forward. Light 'rays' from distant stars are 'tilted' in a similar fashion by the Earth's motion through space. The effect is called aberration.

Many astronomers took up Bradley's techniques, notable among them Friedrich Bessel, a German born in 1784, who catalogued the positions of thirty thousand stars and was one of three astronomers who each independently cracked the parallax problem at about the same time, in the late 1830s. William Herschel had been one of those who tried and failed to measure parallax in stars – sidereal parallax. He tried a trick which would have done away with the need for absolute precision in the catalogues, looking at pairs of stars very close together on the sky, in the hope that one might be very far away along the line of sight while the other might be close enough to show a parallactic displacement. If such a pair could be found, the displacement would only have to be measured relative to the more distant star, not to any absolute standard. But instead Herschel found that many of the double stars he looked at really were doubles, genuinely close to each other in space, orbiting each other like the

Moon and the Earth. That was an important discovery, but not what Herschel had been looking for.

The breakthrough came when the observations were good enough to measure the tiny displacements of the stars involved, and when theories were good enough to eliminate all of the other factors, such as aberration and nutation, that also made the positions of the stars change with the seasons. Success came when the time was ripe, not before; but when the time *was* ripe, success came in a rush. The three-pronged attack of the 1830s came from Bessel, who chose the star 61 Cygni to study because it has a large proper motion, 5.2 seconds of arc a year, and therefore must be very close; from Thomas Henderson, a Scot, born in 1798, who was working in South Africa and chose to study Alpha Centauri, the third brightest star in the night sky, on the grounds that it must be close to look so bright; and from Friedrich von Struve, born in 1793, a German working in Russia, who chose to study Vega (also known as Alpha Lyrae), the fourth brightest star in the night sky, for the same reason. Bessel was the first to announce a successful conclusion to his work, late in 1838; Henderson actually completed his crucial set of observations first, but only announced his findings when he returned to Britain in January 1839; von Struve's measurements, the icing on the cake, appeared in 1840. The three parallaxes they had discovered were indeed small – 0.3136 seconds of arc for 61 Cygni, 0.2613 seconds of arc for Alpha Lyrae, and 1 second of arc (later refined to 0.76 seconds of arc) for Alpha Centauri. The Alpha Centauri parallax is the largest known; the star (actually three stars orbiting one another) is the closest companion to our Solar System, 1.3 parsecs, or 4.3 light years, away from us. Alpha Lyrae, Vega, is 8.3 parsecs (27 light years) away, and 61 Cygni, which is now known to be a double star, is at a distance of 3.4 parsecs, some 11 light years. For the first time, astronomers had a true grasp of just how isolated the Solar System is in the dark emptiness of space. The nearest star is seven thousand times further from the Sun than its most distant planet known today, Pluto. And once they knew the distances to even a few stars, astronomers could work out their true brightnesses, and thereby get a rough idea of the distance to faint stars too remote to show any measurable parallax displacement. With this, and other techniques, astronomers in the second half of the nineteenth century at last began to comprehend, in numerical terms, the size and shape of our own Milky Way Galaxy. But it was only in the twentieth century that they were first able to expand the parallax technique to encompass a large number of stars, and then to move on to the realm of the nebulae.

| | Distance | |
Star	Light-years	Parsecs
Alpha Centauri	4.29	1.32
Barnard's star	5.97	1.84
Wolf 359	7.74	2.38
Sirius	8.7	2.67
61 Cygni	11.1	3.42
Procyon	11.3	3.48

Table 3.1 Distances to some of the nearby stars.

Stepping stones to our Galaxy

It took another sixty years for even the parallax method to make much progress, because it was only then that the use of photographic plates on the end of a telescope, instead of human eyes, became standard astronomical practice. A photograph has two key advantages over the eye. First, of course, it provides a direct, permanent record of star positions, which can then be studied at leisure and measured accurately, even using a microscope to gauge precisely the positions of the star images relative to one another. Second, unlike the human eye, a photographic plate, or film, can 'see' very faint objects. The longer you leave the plate exposed, the more light falls on it, and the stronger each faint image becomes; with the human eye, no matter how long you stare into space you won't see anything fainter than you could see when you started to look. So astronomical photography provided many more stars to study, and made it possible to measure the positions of each of them more accurately. In 1900, when the technique was introduced, parallaxes had been measured for just sixty stars. Half a century later, in 1950, the number of known stellar distances was close to ten thousand; but not all of these distances were direct parallax measurements.

Three techniques, in particular, gave astronomers stepping stones from the tiny volume of space around the Sun in which parallax measurements provide a reliable guide to distances – only out to about 30 parsecs, or 100 light years – into the Galaxy at large. Although it quickly became clear that not all stars have exactly the same brightness, spectroscopic studies of the stars, following Huggins's pioneering work, showed that there are family resemblances. A star which has a particular pattern of lines in its spectrum might be identified and its distance measured by parallax, so that its true brightness was known. Then, when another, more distant star was found to have the same spectral type, it was a reasonable guess that it had the same intrinsic brightness. This more distant star appeared fainter, of course, and by measuring precisely its brightness (or faintness) compared with the star whose distance was known, the distance to the more remote star could be estimated.

The other two techniques depended on geometrical tricks, but also on spectroscopy. The crucial spectroscopic element, which we shall find of even greater importance to the story of galaxies, is the shift in the position of those characteristic sharp lines in the spectrum when the source of light producing the spectrum moves towards or away from us. Think first of an object moving away from us. Any waves it emits – whether they are light waves in the case of a star, or sound waves in the case of an object, on Earth, such as a police car – are stretched out by the motion. Stretching a wave makes its wavelength longer; in the case of sound, it makes the note deeper; for light, it shifts the wavelength of visible light towards the red end of the spectrum.[2] When the source of the wave is moving towards us, the waves are squashed, bunched up together to make a more high-pitched noise, or to shift the light towards the blue end of the spectrum.

The discovery that the observed frequency of a sound wave depends on the velocity of the source relative to the observer was made in 1842 by the Austrian physicist Christian Doppler; it is called the Doppler effect, in his honour. Doppler himself realized that similar changes would affect light from a moving source, and in 1848 the French physicist Armand Fizeau gave the first clear description of this redshift or blueshift effect.

The important thing is that the amount of the shift depends on the speed with which a star is approaching or receding from us. Because the whole spectrum is squeezed or stretched bodily, the wavelengths at which those characteristic lines, like the sodium lines, appear are shifted to the red or to the blue by an amount which depends on the velocity of the source along the line of sight. So by measuring the exact positions in the spectrum of a star at which familiar lines appear, and comparing these with the wavelengths at which the same lines show up in light from a suitable source in a laboratory on Earth, astronomers can infer whether a star is moving towards or away from us, and at what rate. This tells you *only* about motion along the line of sight, of course. A star may well be moving across the line of sight, with a transverse velocity, as well. Its actual motion through space will be at some angle to the line of sight, and this actual motion can be found by adding together, geometrically, the two velocities found by observations: the transverse velocity, or proper motion, and the velocity along the line of sight, determined by spectroscopy, the redshift or the blueshift.

So how can we interpret these measured velocities in terms of distances? One trick works only for clusters of stars, groups that are moving together through space, which are not too far away from the Sun. A group of stars all moving in the same direction are, in effect, running along parallel lines, like a railroad track. And, just as the lines of a railroad track seem to converge on a point in the distance, so the motions of stars in such a group, measured proper motions determined over years of observations, will seem to converge on a point in the sky, provided the cluster is relatively close by. This has the great advantage of telling astronomers in which direction through space – at what angle to the line of sight – the cluster is moving. So, when they measure the Doppler shift for the stars in this cluster, they get not only a measure of the velocity along the line of sight, they also know, from the angle the stars' true motion makes to that line of sight, what proportion of its overall velocity this represents. The proportion that is left over must be the stars' true velocity, in kilometres

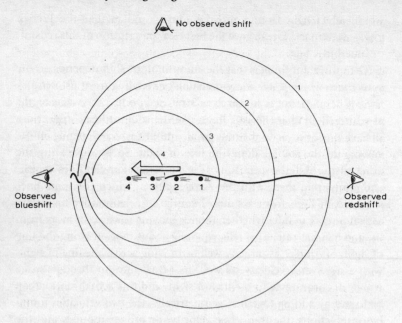

Figure 3.3 The Doppler effect compresses light waves from an object approaching the observer and stretches light from a receding object. The circles represent light emitted from this particular object at the points labelled. Although light travels out at the same speed in all directions, the circles are bunched together in the direction of motion.

per second, *across* the line of sight. And since we already know the proper motion in terms of motion across the sky in seconds of arc, bingo! We can construct one of those imaginary skinny triangles to deduce just how far away that particular cluster must be, in order for *that* speed in kilometres per hour to produce *this* shift in seconds of arc per year. It is a lovely trick, and although it still only works for clusters of stars that are within a few tens of parsecs of the Sun, it did enable astronomers to work out the distance to one such cluster in particular, the Hyades cluster, which turned out to contain a lot of different types of star, all at a distance of about 40 pc. This enabled them to calibrate the brightnesses of those different spectral types, and thereby to use the brightness technique on whole families of stars too remote to show measurable motion across the sky.

The other important geometrical technique sounds almost too naïve to work, but it does. Take a whole lot of stars, just close enough to us for their proper motions to be measured. They might be in the same general direction in space, as viewed from Earth, or they could be scattered all over the sky, chosen, perhaps, on the basis that they all have the same colour, or the same kind of spectrum. Some will be moving this way, some that; some will be moving faster than others, some relatively slowly. But the Galaxy as a whole – certainly the region around the Sun – doesn't seem to be either collapsing in upon itself or exploding outwards. The stars, by and large, are orderly. So it must be that on average all of these random motions cancel out. *On average*, we guess, there is as much chance of a star going one way as another. So, if we add up the velocities of all the stars in our group along the line of sight and take the average Doppler velocity, then we would expect the average velocity of the same group of stars in any other direction, and in particular the velocity *across* the line of sight, to be much the same. Assuming this is so, it is possible to assign an 'average distance' to all the stars in the chosen group by comparing the presumed actual velocities across the line of sight with the measured angular proper motions.

The technique, called 'statistical parallax', is pretty hopeless if applied to an individual star, but the more stars you have to play with, the more reliable the average becomes, so it isn't too bad as an indicator of the distances to some stars – crucially, it proved possible, using this trick, to get a rough idea of the 'average distance' to a group of stars that includes a couple of one particular type, the Cepheid variables. Those distances provided a yardstick for the whole Milky Way system, and out into the Universe beyond. Our scale of the Universe, as I am about to explain, depends on knowing the distance to one or two Cepheids. There are now other techniques, which I won't go into in detail, which have improved those first estimates of the Cepheid yardstick. They depend on the colours of stars and their apparent brightnesses. It turns out that if you take a group of stars that really are physically associated in a cluster the colours of individual stars can be plotted on a graph called a colour-magnitude diagram.

The position of the line made by such a plot depends on the brightness of the cluster, as seen from Earth, and this position can be adjusted so that all such clusters fit on to a standard line, provided that allowance is made for the distance to each cluster. In other words, assuming the stars in each cluster operate on the same physical principles (and if that isn't true we can forget about trying to do astronomy at all) we can fit each cluster in place on the colour-magnitude diagram by assigning it a unique distance from us. But still we have to know the distance to at least *one* cluster, from one of the parallax techniques, in order to calibrate the distance scale on the colour-magnitude diagram in the first place.

Of course, astronomers could see, and investigate, objects much further away than they could measure by parallax. But they could only *guess* just how far away those more distant objects might be, and they could only estimate the true extent of the Milky Way Galaxy. No wonder that the idea of other galaxies, far beyond the Milky Way system, seemed to many astronomers faintly ridiculous. Until, that is, a new yardstick of measurement turned up – a yardstick that could stretch space to give a direct measurement of the distances to some of those extragalactic nebulae.

The Cepheid yardstick

Just as the first steps out from the Solar System into the Milky Way depended on finding distances to the nearest stars, so the first steps out from the Milky Way Galaxy into the Universe at large depended on finding distances to our nearest neighbours in extragalactic space, two nebulae called the Magellanic Clouds, which are visible in the southern hemisphere skies. They were named after the explorer Magellan, having been described by the official chronicler on his circumnavigation of the globe in 1521. This was the first that European civilization knew of the clouds, one large and one small, which look like pieces of the Milky Way that have been broken off. In the sixteenth century, nobody knew what the clouds – or the Milky Way itself, for that matter – might be. Indeed, they were largely ignored by astronomers until John Herschel, the son of William, carried out his survey of southern hemisphere stars and nebulae in the 1830s. By the beginning of the twentieth century, there was no doubt that these clouds, like the Milky Way, were collections of stars. But the idea of nebulae as island galaxies had gone out of fashion, and astronomers generally felt that the Magellanic Clouds were part of the Milky Way system, or perhaps very small semi-independent systems only just outside the Milky Way Galaxy, minor satellites tied to it by gravitational apron strings. The truth about the Magellanic Clouds – and about the scale of the Universe – emerged not through some blinding flash of inspiration, or the observation of some new phenomenon, but as one product of a painstaking and meticulous cataloguing and analysis of thousands upon thousands of stars begun by Edward Pickering at the Harvard College Observatory in the last quarter of the nineteenth century.

Pickering was born in Boston, Massachusetts, in 1846. He taught physics at the new Massachusetts Institute of Technology in the 1860s and 1870s, and was appointed Professor of Astronomy and Director of the observatory at Harvard in 1876. Over the next four decades, he was responsible for several new catalogues of the heavens, each bigger and better than its predecessor, and he was also the inspiration for a whole generation of astronomers. In keeping with the spirit of the times, the tedious job of cataloguing the positions and brightnesses of stars by filling out long rows of figures meticulously neatly written in black ink went to underpaid women; less characteristically of the period, Pickering allowed, and then encouraged, a few of those women to move on to higher things, gaining a toehold in the almost exclusively masculine academic world of the time. One of these women was Henrietta Swan Leavitt, who was given the task of identifying variable stars from photographic plates of the southern sky, obtained by Pickering's brother William at an observing station in Peru.

Henrietta Leavitt was born in 1868 (on the fourth of July; a real Daughter of the Revolution) and studied at the Society for the Collegiate Instruction of Women, which later became Radcliffe College. She joined Pickering's programme at the Harvard College Observatory in 1895 as a volunteer research assistant, received a permanent post in 1902, and soon became head of a department there. Pickering was, no doubt, delighted to have on his team someone with the skill, patience and ability needed to make some sense out of the stacks of photographic plates from Peru, even though neither of them could have had any inkling in 1895 of what was to come out of Leavitt's research over the next seventeen years.

Variable stars – stars which vary in brightness – are obviously of interest to astronomers. Most stars seem to stay the same, at least over a human lifetime, and anything out of the ordinary is invariably a focus of attention. Some variables are really two stars, orbiting round one another so that each in turn eclipses its companion and conceals the light from it. Others, we now know, are stars which pulsate, swelling up and then shrinking in upon themselves, repeating the process over a regular cycle during which their light waxes and wanes. The Cepheids are like this. And some – a few – stars are violently variable, exploding outwards in a brief surge of energy, after a lifetime of quiet normality, before they collapse and fade away into stellar cinders. One of the great advantages of astronomical photography is that, by comparing photographs taken days, months or years apart, it is possible to identify all these different kinds of activity. You can even investigate phenomena that weren't known to be important when the photographs were taken. In the course of her work, Leavitt identified 2,400 variable stars (half of the total known to astronomy at the time of her death in 1921), as well as four of the exploding stars, called novae. And it was her study of one particular kind of variable star that gave her the key to the Universe.

The family of variable stars called Cepheids gets its name from Delta Cephei, which was identified as a variable by the young English astronomer John Goodricke in 1784. He died just two years later, at the age of twenty-one. Cepheids show a characteristically regular pattern of variation in brightness, but different Cepheids have different periods for this variation, some less than two days, others more than a hundred days. The average is about five days. They can be identified as members of the same family, however, both by the typical way in which they brighten and dim, and because they show a family resemblance in their spectra. One of the interesting questions about Cepheids, of course, is why there should be such a range of different periods, when each individual star shows a constant periodicity. As Leavitt continued her painstaking work, identifying the Cepheids (and other variables) on the photographic plates, and noting down the length of each one's cycle and its average apparent brightness, she began to see a pattern emerging. The brighter a Cepheid was, the more slowly it went through its cycle of variation.

In 1908, Leavitt said as much when she published a preliminary report on the progress of her work. It took another four years for this impression to be pinned down in numbers in black and white. But when it was pinned down, in 1912, it provided real hope of establishing an accurate distance scale for the Galaxy, and it was all thanks to the Magellanic Clouds.

At that time, Leavitt had identified twenty-five Cepheids in the smaller of the two Magellanic Clouds (sometimes called the SMC, for obvious reasons). These very clearly showed the relationship between brightness and period, which hardly shows up at all for variables in the Milky Way itself. The reason is easy to see. Stars in the Milky Way are scattered at many different distances from us. Some are nearby, others ten, or a hundred, or even more, times further away. A star that is twice as bright and twice as far away as another star actually looks the fainter of the two – apparent brightness depends on the actual luminosity divided by the square of the distance. So the period-luminosity relation, as it is known, was masked by distance effects within our Galaxy.

But things are different for the stars in the Small Magellanic Cloud. The cloud is so far away from us that all of the stars in it can be regarded as being roughly the same distance from Earth. One may be a little closer than another, but not one of them is even as much as twice as far away as any of the others. The scale of the differences in distance is much less than the average distance to the cloud, in the same way that to me, writing this book in a small village in England, everybody in New York can be thought of as the same distance away. The nearest town to me is a little over a mile away, and the far side of the town is more than twice as far away as the nearest side. A difference of a mile is important when I am planning a journey into one part of town or another. For all practical purposes, however, I am equally distant from Times Square and from the Statue of Liberty. A couple of miles counts for little compared with the width of the North Atlantic Ocean.

So it was that Leavitt was able to work out the relationship between luminosity and period for Cepheids using her twenty-five variables in the SMC. She found that, for example, if one Cepheid has a period of three days and another one of thirty days, then the star with the longer period is six times brighter than the star with the shorter period. Assuming the rule she found for Cepheids in the Magellanic Clouds holds for all Cepheids, this immediately meant that Cepheids in the Milky Way could be used to give an indication of the *relative* distances to stars and clusters of stars across the Milky Way. But nobody knew the actual intrinsic brightness of even one Cepheid, so the distance scale was uncalibrated. Astronomers had a measuring stick for the Galaxy, but they didn't know the length of the stick; they could tell that one star, or cluster of stars, was twice as far away as another, but they didn't know if they were measuring, to mix the analogy, in miles or kilometres. So they still didn't know if the Magellanic Clouds were small systems within the Milky Way, or much more distant objects, galaxies in their own right.

It only took a year for the truth to emerge. Ejnar Hertzsprung, a Danish astronomer and physicist (who was born in 1873 and remained active in research until just before his death in 1967), made the first estimate of the distances to some of the nearer Cepheids, using a variation on the statistical parallax technique. With all its imperfections, that technique gave him an indication of the actual distances to one or two Cepheids. By comparing their apparent brightnesses with the distances, he could easily calculate the actual brightness in each case. From that it was a simple step to calculate the actual distance to any other Cepheid, using its period to indicate how much brighter or dimmer it must be in reality compared to the few whose distances and absolute luminosities had been measured. Hertzsprung concluded that the SMC was 30,000 light years (in round terms, 10,000 parsecs) away, much further than any one had suspected. But this measurement did not immediately open the eyes of astronomers to the true size of the cosmos, for two reasons. First, because he had not allowed for the fact that dust in space blocks out some of the light from distant stars and makes them look dimmer than they really are, Hertzsprung's calibration was a little off – the best modern calculations give an even more impressive distance of 170,000 light years, or 52,000 parsecs, to the Large Magellanic Cloud, and 63,000 parsecs (63 kiloparsecs, kpc) to the SMC. And, second, astronomers were far too busy using their wonderful new yardstick to measure the size of the Milky Way to worry much, for the next few years, about what lay beyond the Milky Way. To go beyond the Magellanic Clouds and into the real Universe required a new leap of the imagination, and a new generation of telescopes. Before we make those two leaps, however, it is only right to acknowledge the great achievements of the astronomers who did map out our own Galaxy, using techniques that form the foundations of later investigations deeper into space.

The scale of the Galaxy

The two men who were together largely responsible for the next step towards an understanding of the scale of the Universe came from very different backgrounds. George Ellery Hale, the greatest telescope builder of the twentieth century – perhaps of all time, even allowing for the advance of technology since the time of Galileo, Herschel or Rosse – was a wealthy man, the son of an elevator manufacturer. He was born in 1868, in Chicago, and his education progressed smoothly through conventional channels to MIT and an appointment, in 1892, as Professor of Astronomy at the University of Chicago. Hale's enthusiasm for astronomy had been fired in childhood when he learned that light from the Sun could be analysed by spectroscopy to reveal the composition of our nearest star. By the age of twenty, he was designing new kinds of spectroscopic instruments with which to dissect the Sun's light more efficiently; it was a lifelong dream of this astronomer, born ten years after the publication of Darwin's *Origin of Species*, that science might one day be able to explain the origin and evolution of stars and the origin and evolution of life in one grand package. Today, it is possible to argue that that dream is all but fulfilled – that is the justification for writing this book. And it is in no small measure due to Hale's enthusiasm and skill as a telescope builder, fund raiser and observatory director that we are so close to fulfilling his dream.

Hale's career as the moving force behind the construction of a new generation of telescopes and observatories began by chance, when he heard that the University of Southern California had ordered the lenses to make a 40-inch refracting telescope, but had been unable to pay for them. Telescopes are measured in terms of the diameter of their main magnifying lens or mirror, and at that time, in the 1890s, the largest refracting telescope (that is, one using a main lens rather than a mirror) was the 36-inch at the Lick Observatory, on Mount Hamilton near San Jose in California. This is close to the practical limit for constructing accurate astronomical lenses, because bigger ones are bent out of shape by their own weight. The biggest telescopes of today are all reflectors, using big parabolic mirrors, not lenses, to focus the light they gather from the stars. A mirror has the great advantage over a lens that, because no light is passing through, the back can be supported by a framework to hold the mirror in shape. By the 1890s, it was beginning to be clear that the next step forward in telescope design would involve large mirrors; but Hale was intrigued by the possibility of obtaining these ready-made lenses, which were in store at the Paris workshop where they had been manufactured, and using them to build an even bigger telescope than the one at Lick. As the son of a wealthy man, he had the right connections to seek the money needed for the project, and duly made the rounds of other wealthy Chicago families, eventually obtaining a promise of the required funds from Charles Yerkes, a trolley-car magnate. The sum required was $349,000, which Yerkes only grudgingly coughed up, in dribs and drabs, over the next few years, as Hale kept up his campaign. But the money did come, and the telescope was built, forming the centre piece of the Yerkes Observatory of the University of Chicago, with Hale appointed as its first Director, in 1897, at the age of twenty-nine. The 40-inch (roughly 1-metre) Yerkes telescope is still the largest refractor in the world.

Hale now had the bit between his teeth. The big refractor, in its observatory at Williams Bay, Wisconsin, was all very well, but he wanted something bigger and better, located in an even more desirable site. The best place to see the stars from Earth is on the top of a high mountain, clear of all the dust and clouds in the lower atmosphere of our planet, and far away from city lights. Soon, Hale was off to Mount Wilson, in California, camping out in an abandoned shack while testing the view of the heavens using a small telescope. Back on the stump again, he drummed up support from the Carnegie Institution of Washington for a new observatory to be built on Mount Wilson, equipped initially with a reflecting telescope with a main mirror 60 inches (1.5 m) across, and with Hale as its Director. The mirror itself was a gift from Hale's father; the telescope came into use in 1908, and was the main tool used by the man who established the true scale of the Milky Way, our Galaxy.

Harlow Shapley came from a farming background, and was born in Missouri in 1885. He received little formal education as a child, and at the age of sixteen he was working as a crime reporter on a Kansas newspaper. But Shapley figured that a formal education would help his career, and after two years at the Carthage Presbyterian Collegiate Institute, he set off to the University of Missouri to enrol in the journalism course. On arrival, in 1907, he found that the course would not open for another year, and, feeling that he had wasted enough time already trying to catch up on his education, he decided to study something else – anything else – rather than hang about. Late in life (Shapley died only in 1972) he always said that he picked astronomy because it began with the letter 'A', and so caught his eye near the top of the list of courses available. After four years, Shapley emerged with both BA and MA in his randomly chosen speciality, and went on to Princeton, where Henry Norris Russell set him the task of studying binary stars. Most of Shapley's work over the next three years concerned eclipsing binaries; he also established once and for all that the Cepheid variables are not binaries, but are pulsating stars. And in 1914, seven years after finding that the University of Missouri had no journalism course to offer him, Shapley emerged from Princeton with a PhD and a reputation as one of the brightest of the new generation of astronomers. His reward was a job at the new observatory on Mount Wilson, with a salary of $135 a month and, much more important, access to the biggest telescope in the world, the new 60-inch.

This was one year after Hertzprung had made the first attempt to use Cepheids as distance indicators, and Cepheids had formed a part, albeit the lesser part, of Shapley's own PhD work. With the best telescope in the world at his disposal, Shapley set out to map the Milky Way, using Cepheids. The approach he used picked out another feature of our Galaxy, something quite different from anything we have encountered so far. These are the globular clusters, spherical groups of stars which each contain anything from a few tens of thousands up to a few tens of millions of separate stars packed together, which shine like beautiful jewels in the field of view of even a modest telescope.[3] These globular clusters lie mainly in one part of the sky, and they seem to be arranged in a sphere themselves. But was it a small sphere, close by, or a large sphere, far away? Fortunately, globular clusters often contain regular variables – several may be found in just one cluster. So Shapley was able to use the new yardstick, and the new telescope, to find out the true distances to some of the clusters. As he did this, he found that the brightest stars in each cluster always seemed to be about the same intrinsic brightness as the brightest stars in any other cluster. With the aid of the Cepheid yardstick, he had found a new measure of the Milky Way – the distances to even those globular clusters where no Cepheids could be seen could be estimated by assuming their brightest stars were the same brightness as those in other clusters, and calculating distance from the apparent brightness (or dimness) of those giant stars.

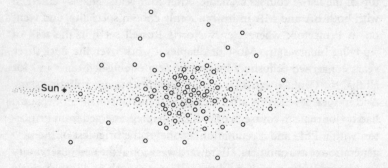

Figure 3.4 Harlow Shapley found that the distribution of globular clusters, represented here by circles, shows that the Sun and our Solar System lie far from the centre of the Milky Way system.

The outcome of all this was a new map of our Galaxy. The globular clusters were seen to fill a vast sphere very far from us, centred on a point in the direction of the constellation Sagittarius, at the heart of the Milky Way. The only plausible conclusion was that the centre of this spherical system is indeed the centre of the Milky Way Galaxy, and that the Sun and Solar System are located well out in the stellar suburbs, about two-thirds of the way to the edge of the spiral system of stars. Shapley's results were published in a series of papers over a period of several months in the years 1918 and 1919. The actual size he came up with for the whole Milky Way system was nearly three times too big, because he had not allowed for the effects of obscuring dust on the light from distant globular clusters (a cluster whose light is dimmed by dust looks fainter, and so Shapley estimated it as being further away than it really is). But his main conclusions were correct. We now calculate that the Milky Way Galaxy is a flattened disc, just as Wright and Kant imagined, about 100,000 light years (30 kiloparsecs) across. The distance from the centre of the disc to the edge is about 50,000 light years, and the Sun is about 30,000 light years (10 kpc) out from the centre. We live very much in the backwoods of our own Galaxy. But how significant is our Galaxy in the cosmic scheme of things? Leavitt, Hertzprung and Shapley had together pushed astronomy's best estimates of how far 'up' we can see into the sky out into the hundreds of thousands of light-years' range. How much further might 'up' extend?

Across the Universe

At this point Shapley faltered and took a wrong turning. It wasn't entirely his fault – his attempt to provide a mental picture, an imaginary model, of the whole Universe depended on observations and interpretations made by others. But the mistake was to set his whole career on a different path, so that although he achieved considerable eminence and respect among his peers he always looked back on the years from 1914 to 1920, on Mount Wilson, as the pinnacle of his career.

Shapley's inflated estimates of the size of the Milky Way made it look as if the Magellanic Clouds were just part of our own Galaxy, not separate galaxies in their own right. And if the clouds were not real galaxies, the obvious conclusion, for Shapley, seemed to be that other nebulae, such as the great spiral in the constellation Andromeda, must also be, at best, minor satellites of our own Galaxy. According to this picture, the Milky Way *was*, essentially, the Universe; the limit of 'up' had already been seen. But there were other astronomers who thought that the nebulae must also be galaxies in their own right, and that Shapley's estimate of the size of the Milky Way must be too big, even if they didn't know why it was too big. This alternate view was expressed most vociferously by Heber Curtis, an astronomer at the Lick Observatory.[4]

Curtis was another researcher who came to astronomy by a strange route. Born in 1872, in Muskegan, Michigan, he studied classics and became Professor of Latin at Napa College, California, at the age of 22. There, he became interested in astronomy, and when the college merged with the University of the Pacific in 1897 he became Professor of Astronomy and Mathematics – a somewhat startling turnaround to anyone used to the career structure in universities today! After several short periods of research at various observatories, Curtis settled at Lick in 1902 where he stayed, apart from a spell observing the southern skies from Chile, until 1920. It was after he returned from Chile, in 1909, that he concentrated on determining the nature of what were then still known as the spiral nebulae. Increasingly better photographs of these nebulae were becoming available in those years, and it was these pictures that convinced Curtis that the nebulae were galaxies like our own, sometimes seen edge on, sometimes viewed in plan, with the full glory of the disc and spiral structure visible, and sometimes seen at intermediate orientations. If so, these external galaxies, beyond the Milky Way, must lie at truly vast distances in order to appear only as little clouds of light in our telescopes. But how could the distances to the nebulae be measured? That question was open to two answers in the years up to 1920, and one of the interpretations was based on an unfortunate misunderstanding of an event that was seen on Earth in 1885. Unfortunately for Shapley, he backed the wrong horse.

It was on 20 August 1885, that Ernst Hartwig noticed a new star, or nova, shining in the Andromeda nebula. The star soon faded, but not before its peak intensity had been noted. This was the first time an individual star had been seen associated with the nebula, and one interpretation of the event was that Hartwig and his contemporaries had witnessed the birth of a new star out of a swirling cloud of gas and dust within the Milky Way system. Whatever the star was, it had briefly shone as bright as all of the rest of the Andromeda Nebula put together. A chance to calibrate just how bright that might be came, it seemed, in 1901, when another star exploded, in the direction of the constellation Perseus. This nova was so close that its distance could be estimated, by parallax, as about 100 light years from the Sun. With no better estimate of distances available, astronomers of the time guessed that the nova in Andromeda might have been the same absolute brightness as the nova in Perseus, and that would have placed it, judging by its apparent brightness, sixteen times further away, just 1,600 light years from us. That meant that the Andromeda Nebula must be quite large, but still placed it within the Milky Way system.

This was, essentially, the reasoning adopted by Shapley in support of his claim that the Milky Way dominated the Universe and that the spiral nebulae were mere incidentals. Curtis, convinced that the nebulae were galaxies in their own right, sought evidence to back his case up. Suppose that the new star seen in Andromeda in 1885 really had been much brighter than the nova in Perseus in 1901. If the Andromeda Nebula were a whole galaxy like the Milky Way, it would, indeed, have had to be as bright, briefly, as a thousand million ordinary stars, something which struck Shapley as absurd. Who can blame him – but we now know that very occasional 'supernovae' do indeed shine that brightly. And one reason we know that is because Curtis set out to find other novae in Andromeda and to compare their brightness with the event of 1885 and the brightness of Nova Persei in 1901.

The fact that Curtis actually found several novae in Andromeda (more than a hundred have now been recorded) proved that it must be a collection of very many stars, since novae are not all that common. And the fact that all of those novae were much dimmer than the event seen in 1885 suggested that these were the right phenomena to compare with Nova Persei. That revision increased the distance scale to Andromeda more than a hundredfold, placing it hundreds of thousands of light years away, far beyond the edge of the Milky Way. So who was right, Curtis or Shapley? There was so much interest in the issue that the National Academy of Sciences organized a debate between the two astronomers in Washington, DC, in 1920. The debate, attended by, among others, Albert Einstein, was widely publicized. The general feeling was that Shapley lost – Curtis's interpretation of the scale of the Universe was correct.[5] It was in the wake of this defeat that Shapley quit Mount Wilson, heading off to Harvard, where he took up the directorship of Harvard College Observatory, which he had first been offered in 1919, when Pickering died. It was a decision he must have regretted, in spite of his many other contributions to astronomy, as he saw a new man at Mount Wilson, Edwin Hubble, pick up where he had left off. Hubble built on Shapley's technique of estimating distances using Cepheids and globular clusters; and he also had a new telescope to play with, even bigger and better than the 60-inch.

Hubble was the greatest single beneficiary of Hale's industrious activity. Not content with just the 60-inch telescope for his Mount Wilson Observatory, Hale had cajoled a Los Angeles businessman, John D. Hooker, into funding a telescope with a main mirror 100 inches (2.5 metres) across. This Hooker telescope was completed in 1918, and was for three decades the biggest telescope in the world. Hale himself, worn out by his efforts, resigned as Director of the Mount Wilson Observatory on medical advice in 1923, when he was fifty-five. His idea of a quiet retirement, at his home near Pasadena, involved building a small observatory and inventing a new kind of spectroscope to study the Sun. He then tried to raise money to build an observatory in the southern hemisphere and failed, suffering his second nervous breakdown. But soon he was back in circulation again with another scheme, for yet another giant telescope, this one to have a mirror 200 inches (5 metres) across. The Rockefeller Foundation came up with $6 million for the project, to be carried out under the auspices of the California Institute of Technology, in 1929. Hale was chairman of the group planning the construction of his masterpiece, which was to be built on Mount Palomar in California. The project took twenty years to complete, delayed by, among other things, World War II, and Hale died in 1938 before seeing his masterwork finished.[6] It came into operation, as the Hale telescope, in 1948; in 1969 the twin observatories on Mount Wilson and Mount Palomar were, fittingly, renamed the Hale Observatories in tribute to the man who placed

American astronomy at the forefront in the twentieth century. By then, Ed Hubble had long since opened the eyes of all astronomers to the true scale of the Universe.

Hubble, like Shapley, was born (in 1889) in Missouri. He came from the town of Marshfield, the fifth of seven sons of a local lawyer. He attended both high school and university in Chicago, overlapping at the University with Hale's time there as a professor. Hubble was a superb natural athlete, and was offered a chance to turn professional as a boxer to fight the great Jack Johnson. Instead, he took up the offer of a Rhodes scholarship to travel to Oxford University in England, where he studied law, represented the University as an athlete, and fought the French boxer and champion Georges Carpentier as an amateur in an exhibition bout. On his return to the United States in 1913 Hubble joined the Kentucky bar, but practiced as a lawyer for only a few months before deciding this was not the career he wanted. Reverting to an interest in astronomy that had been partly stimulated by Hale during his University of Chicago days, Hubble returned to that university, studying astronomy and working as a research assistant at the Yerkes Observatory. He finished this research in 1917 and was awarded a PhD; Hale offered him a post at Mount Wilson, but first Hubble enlisted in the infantry and went off to fight in France, where he was wounded by shell fragments in his right arm. So it was in 1919 that he arrived at Mount Wilson, just as the new 100-inch telescope was coming into full use, and just before Shapley departed for Harvard. His timing couldn't have been better. It had only been as recently as 1917 that a nova had been identified for the first time on a photographic plate (by George Ritchey, at Mount Wilson), stimulating Curtis to search back through the photographic records at Lick and find the evidence which gave him the first direct measure of the distances to extragalactic nebulae. For hundreds of years, the nature of those nebulae had been open to debate; by 1924, the debate was over, and the combination of the 100-inch telescope and Edwin Hubble had given mankind a new picture of the Universe, with more startling discoveries still to come.

Hubble's Universe

Hubble believed that the spiral nebulae were galaxies, far beyond the Milky Way, but he wasn't going to be rushed into any overhasty attempt to prove the point. First he tackled the problem of the other nebulae, the ones that didn't show the characteristic spiral structure and which were almost certainly part of the Milky Way system. Using a variety of telescopes, often the 60-inch and occasionally, at first, being allowed time on the 100-inch, by 1922 Hubble had completed a major study which showed that these gaseous nebulae (they also contain dust) shine not with their own light like stars but either because they reflect the light from stars within or close to the nebula, or because the energy the nebula absorbs from nearby stars is enough to make the hot gas glow. The association between gaseous nebulae and stars in our Galaxy was confirmation that the nebulae themselves were indeed part of the Milky Way system. But what of the spiral nebulae? His 'apprenticeship' served, Hubble now turned his attention to the problem closest to his heart.

Even in the early 1920s, and even with the aid of the 100-inch telescope, it still had not proved possible to obtain pictures of any spiral nebulae that clearly showed them resolved into separate stars, like the Magellanic Clouds. The best photographs Hubble could obtain seemed, under a magnifying glass, if the light was right and Hubble's mood was optimistic, to show a hint that the wash of light might be broken up into the granular structure that would reveal the nebula in question to be a collection of individual stars. But it wasn't the sort of evidence that as cautious a man as Hubble would stake his reputation on. If the spirals couldn't be resolved into separate stars, Hubble determined to investigate a star cloud that *could* be resolved, even if it was only a faint, irregular patch on the sky, less significant than the Magellanic Clouds. He settled on a group of stars called NGC (for 'New Galactic Catalog') 6822 and spent two years obtaining the best series of photographs of the cloud that he could. On a good night, it might be possible to get one useful photograph of the cloud; on other occasions, it took two separate nights of observing to get a single decent photograph, hour upon hour of patient observing while Hubble kept the cloud locked in the sights of the 100-inch. And, of course, there were other demands on the telescope's time. So it was that it took most of 1923 and 1924 for Hubble to get a set of fifty good plates of NGC 6822. The result was the identification of just over a dozen Cepheids in this cloud and, using Shapley's techniques, Hubble was able to set the distance to this little, irregular galaxy as seven times the distance to the Small Magellanic Cloud. This was in 1924.

In the middle of the observing program on NGC 6822, another extragalactic Cepheid was identified, in the Andromeda Nebula, also known as M31 (number 31 in Messier's catalogue). The discovery was made in the autumn of 1923, during a survey aimed at finding novae in the Andromeda Nebula, novae which might be used to test Curtis's ideas concerning the nature of the nebula. 'The first good plate in the program,' recalled Hubble in his book *The Realm of the Nebulae* (p. 93), 'made with the 100-inch reflector, led to the discovery of two ordinary novae and a faint object which was at first presumed to be another nova. Reference to the long series of plates previously assembled by observers at Mount Wilson in their search for novae, established the faint object as a variable star and readily indicated the nature of the variation. It was a typical Cepheid with a period of about a month . . . the required distance was of the order of 900,000 light years.' For various reasons, that distance estimate has now been revised upwards, to more than 2 million light years (670 kpc). But that is a minor detail compared with the breakthrough of this discovery. With no new assumptions at all (unlike Curtis, who could only *guess* that novae in Andromeda were the same intrinsically as novae in the Milky Way), but using the same yardstick that had been used by Shapley to map the Milky Way, Hubble could now measure the distances to the nearer external galaxies.

That, perhaps, is the most important point to take in just now. The breathtaking leap out to a distance of 2 million light years represented simply the first step out into the cosmos, to one of the *nearest* of the other galaxies like our Milky Way. The whole Galaxy in which we live was suddenly shrunken, in the astronomical imagination, into a tiny mote floating in a vast, dark sea of emptiness.

It took a little while for the actual insignificance of the Milky Way Galaxy to sink in. At first, it seemed that our Galaxy was rather larger and more impressive than the others. It was only in 1952, with a revision of the Cepheid distance scale, that it became clear that other galaxies are just as big as our own, and even further away than Hubble had estimated. With improved photographic emulsions, Hubble succeeded in 1923 in resolving the outer part of the Andromeda Nebula into dense swarms of stars, and identified more Cepheids in both M31 and another spiral about the same distance away, M33, over the next few months and years. Sufficient evidence to settle the issue of the nature of the spiral nebulae was in by the end of 1924, and presented by Hubble to a meeting of the American Astronomical Society; over the next five years Hubble accumulated more evidence, and he produced the definitive summing up of that evidence in the last year of the decade. By then, he had also begun to establish techniques for estimating the distances to nebulae – galaxies – far beyond the range where individual stars could be resolved and the Cepheid yardstick applied.

Cepheids themselves can be identified in only about thirty of the closest galaxies, even using the 200-inch telescope. The Hubble Space Telescope has now improved on this, but other techniques for measuring distances to more remote galaxies will always be needed. The first step Hubble used, again borrowing from Shapley, was to use supergiant stars as distance indicators, just as they are used to indicate the distances to globular clusters in our own Galaxy. That took Hubble out to four times the distance over which Cepheids could be seen, a distance he estimated as about 10 million light years. Globular clusters themselves can be, and have been, used as a rough yardstick of the Universe, on the assumption that the brightest clusters in each galaxy are the same intrinsic brightness as the brightest in our own Galaxy, but by now astronomers are beginning to scrape the bottom of the barrel in their search for ways to measure the distance 'up' to more and more remote galaxies. To go further Hubble had to make a bold, and only roughly accurate, assumption. When he looked at a large cluster of galaxies in the direction of the constellation Virgo, he found that they all appeared roughly as bright as each other – at least, the brightest of these galaxies outshone the dimmest only by a factor of ten. By assuming that all galaxies were equally bright and had an absolute luminosity three times that of the dimmest galaxy, or one-third of that of the brightest, Hubble could estimate distances and be reasonably sure that the estimates would be within a factor of three of the right answers – perhaps three times too big, or three times too small, but no worse than that. This kind of technique was later improved by using only the brightest galaxy in a cluster as the standard; it turns out that the brightest galaxies do seem to be much of a muchness, like the brightest supergiant stars. Approximate though it was, the technique took Hubble out to about 500 *million* light years (these are Hubble's figures, now revised upwards considerably, but they give you a flavour of the advances he was making). That distance encompasses a volume of space that contains about 100 million galaxies. But each of these distance measurements, every one, depends on the initial calibration of the Cepheid yardstick using statistical parallax techniques (and now the colour-magnitude diagram) within our own Galaxy – indeed, within the immediate proximity of the Sun. The seeming wealth of information we now have about distances across the Universe is like an inverted pyramid, expanding upward and outward from that one Cepheid calibration at the point on which the pyramid is balanced. Without measuring the distances to those 30-odd galaxies in which Cepheids have been identified, there is no way to calibrate the other, more rough-and-ready, yardsticks at all. If that Cepheid yardstick has been incorrectly calibrated and is later revised, we have to change the whole scale of the Universe. That, as we shall see, has happened several times over the decades, most importantly in the early 1950s. But none of these revisions alters the basic picture of the Universe established by Hubble.

Hubble's universe – our Universe – extends for hundreds and thousands of millions of light years. Some of the galaxies whose images are viewed today by giant telescopes like the 100-inch and 200-inch are so remote that the light we see them by set out on its journey to us even before the Earth itself was formed. There really is no way in which the human mind can comprehend the size of the Universe. All we can do is gaze at the numbers, which tell us that even our nearer neighbours M33 and M31 are so far away that light takes 2 million years or more to cross the gap from them to our Galaxy, and admit that we are bemused by it all. Even the greatest of cosmologists, an Albert Einstein or a Stephen Hawking, must have some sympathy, in his heart of hearts, with Carlyle's remark: 'I don't pretend to understand the Universe – it's a great deal bigger than I am.'

The only reason that astronomers are able to determine the properties of the Universe at large is that, relatively speaking, galaxies are much closer to one another than stars are. One of the best ways to picture this is by using an imaginary model of the Universe based on aspirins. If our Sun were the size of an aspirin, then the nearest star would be represented by another aspirin 140 kilometres away. This is fairly typical of the spaces between stars – the distance from one typical star to its nearest neighbour is several tens of millions of times the diameter of the star itself (except, of course, for binaries and similar systems where two or more stars orbit closely around one another). Galaxies, like our own Milky Way, contain thousands of millions of stars, spread over appropriately large volumes of space, but all held together, orbiting the galactic centre, by gravity. We can get an idea of the spacing between galaxies by changing our scale so that now the Milky Way, not the Sun, is represented by an aspirin. On this new scale, the nearest galaxy, M31, is represented by another aspirin, just 13 centimetres away.

This is slightly misleading, because both the Milky Way Galaxy and M31 are members of a small group of galaxies, called the Local Group, held together by gravity. The distance to the nearest similar small group of galaxies, the Sculptor Group, is still only 60 centimetres on the aspirin scale, however; and only 3 metres away, on this picture, we find the Virgo Cluster, a huge collection of about two hundred galaxies, spread over the volume of a basketball. The Virgo Cluster is at the centre of a loose swarm of galaxy clusters that it dominates gravitationally; these include both the Local Group and the Sculptor Group, and the whole swarm is known as the Local Supercluster.

We can go on, on this picture. Just 20 metres away there is another big cluster, the Coma Cluster, containing thousands of galaxies. Further out, there are even larger clusters, some 20 metres across. The powerful radio-emitting galaxy Cygnus A is 45 metres distant; the brightest quasar on the night sky, 3C273, 130 metres away. And the entire visible Universe can be contained within a sphere roughly 1 kilometre across, on the scale where an aspirin represents our Galaxy.

It doesn't make much difference which of these distances you choose as representing a typical spacing between galaxies. Even the distance to the Virgo Cluster is only six hundred times the diameter of our Galaxy; M31 is just about twenty-five Milky Way diameters distant from us. If galaxies were as far apart, relatively speaking, as the stars within galaxies, then the distance to our nearest galactic neighbour would be a hundred times further than the most distant object ever seen in the real Universe! Clearly, extragalactic space is far richer in galaxies than galactic space is in stars. And that enables cosmologists to get a broad picture of the way visible matter is distributed through the Universe, and how that distribution has changed as the Universe has evolved.

Astronomers attempt to understand the Universe, or as much of it as they can be aware of. Hubble laid the groundwork for modern cosmology. He not only established the scale of the Universe, he described and classified the main types of galaxies – 75 per cent of those that can be seen are spirals, most of the rest are cigar-, or American football-, shaped ellipticals, and only a few irregulars are seen (probably because most of them are too small and faint to be visible at such distances). And he also analysed the distribution of galaxies through space, and found that the distribution is, by and large, uniform. Although galaxies come in clusters, the clusters them- selves are distributed at random over the sky, and there is just as much chance of seeing a galaxy, or cluster of galaxies, in one part of the sky as in any other, once allowance has been made for the obscuring effects of dust in the Milky Way. This was a major discovery in its own right, suggesting that the ultimate structural pattern of the Universe might have been discovered; its importance is only slightly diminished by very recent evidence that there may be one more layer of structure, involving clusters of clusters of galaxies. It is still fundamentally important that the Universe is the same in all directions, that there is no special place anywhere in the Universe. But even this fundamentally important observation paled into insignificance compared with the bombshell Hubble dropped in 1929. He found that, once we look beyond the cosmic back yard of the Local Group. all of these millions upon millions of galaxies are moving apart from one another, rushing away from each other with speeds of up to a sizable fraction of the speed of light. The whole Universe was found to be expanding; it was that discovery that pointed clearly towards the fact that the whole Universe must have had a definite beginning in time. There seemed to be no limit to how far up astronomers could look into the dark night sky; but the implication of that universal expansion was that there was a limit to how far back in time the history of the Universe extended. It was the discovery of the universal expansion that, only as recently as 1929, scarcely seventy years ago, set astronomers firmly on the trail of the Big Bang.

Chapter 4

The Expanding Universe

Science doesn't always progress in an orderly fashion. A discovery made today may have to wait years, or decades, for its significance to be appreciated and slotted into place, while another observation, made tomorrow, may be of obvious and immediate importance. Different lines of research can proceed seemingly independently for a generation or more, until some linking factor shows them to be simply two facets of a greater whole. The two great lines of research in which Hubble was closely involved weren't quite like that. It was obvious from the moment that the large velocities of galaxies were discovered that these velocities were telling us something significant about the nature of the Universe, just as were Hubble's surveys of the number and distribution of faint and distant galaxies. But progress on both fronts took place in fits and starts, and any historical account of the landmarks in the research has to follow one track for a time, then backtrack to pick up the other main theme.

The work which showed the Universe to be filled with a homogeneous distribution of galaxy clusters, the same in all directions for as far as the telescope can see, continued throughout the 1930s and beyond – it continues, indeed, to the present day. One of Hubble's original ambitions, his colleagues have told many times, was to take a photograph of part of the night sky, using the 100-inch telescope, in which there would be as many galaxies visible on the final print as there were foreground stars from our Milky Way. That ambition was achieved on 8 March 1934, confirming that the galaxies are as numerous in terms of the Universe as stars are in terms of the Milky Way. The date provides a convenient landmark for us, the moment when the extent of the Universe and the nature of galaxies as its fundamental visible building blocks was established beyond reasonable doubt. From his counts of the number of galaxies photographed on plates like this, Hubble calculated that 100 million galaxies were in principle photographable; with the 200-inch and other large telescopes now available, astronomers calculate that a hundred billion (10^{11}) galaxies could now be photographed if we had the time and inclination to survey the whole sky in detail. This is, indeed, in round terms the same as the number of stars in our Milky Way Galaxy. But it was five years earlier, half a decade before he obtained that landmark photograph in 1934, that Hubble had reported the discovery that all these galaxies, except our nearest neighbours, are not only moving away from us as the Universe expands but doing so in accordance with a simple physical law. And it was seventeen years before that, in 1912, that the first measurements of the velocities of what were then still called 'the nebulae' were made.

Redshifts and blueshifts

The story of the redshifts of distant galaxies actually begins with the fascination the red planet, Mars, held for Percival Lowell, the wealthy scion of a prominent Boston family of the nineteenth century. Lowell was born in 1855, and studied mathematics at Harvard University, from which he graduated in 1876. This was a year before the Italian astronomer Giovanni Schiaparelli reported his first detailed observations of *canali* on Mars. The correct translation of *canali* is 'channels'; during his long study of Mars Schiaparelli made it quite clear that he used this term, and the equivalent terms for 'seas' and 'continents', in a purely descriptive sense to identify features on Mars, and not with the intention of making any claim that these features corresponded to the seas, continents and channels (let alone canals) of Earth. But, partly through a mistranslation of *canali* as 'canals', and partly through wishful thinking, Schiaparelli's reports created a wave of interest in France, Britain and North America which lasted for decades, with many serious astronomers, not to mention huge numbers of ordinary folk, convinced that there was intelligent life on Mars, with Martians busily constructing canals to carry water from the polar caps to the equator. The culmination of this misrepresentation of Schiaparelli's observations must, I suppose, have been the famous radio broadcast of Orson Welles's version of H. G. Wells's *War of the Worlds*, which was itself written in the 1890s, at the height of the interest in Mars roused by Schiaparelli. In 1938, this radio play, presented in the form of a factual news account, described an attack on New Jersey by invaders from Mars, and panicked thousands of listeners who didn't realize they were tuning in to a work of fiction.

But all that was half a century in the future when Schiaparelli's discoveries, in their garbled and mistranslated form, reached the United States in the late 1870s and caught the attention of young Percival Lowell. The seed that was planted then was a long time coming to fruition. Lowell spent a year travelling after graduating, then six years in his father's cotton business before quitting to spend the best part of ten years in Japan and the Far East. It was only on his return to the US, in 1893, that he decided to take up astronomy, and especially the study of the planets, seriously. A man of means, he was able to finance the construction of his own observatory at Flagstaff, Arizona, in the clear air more than 2,000 metres above sea level and far from any major cities. For fifteen years he studied Mars using a 24-inch refracting telescope, reporting not only canals but oases and clear signs of vegetation to an eager world. The discoveries owed much to his imagination, but other astronomers made similar mistakes – even the best telescopes on Earth give only a poor image of Mars, because details are blurred by the Earth's atmosphere. A bigger telescope simply magnifies the blurring. Lowell was wrong about life on Mars, but he certainly excited the interest of a generation of Americans in astronomy. And he also predicted that there must be a ninth planet, outside the orbit of Neptune, revealed by its disturbing influence on the orbits of the other outer planets. Pluto was discovered, exactly where Lowell had predicted, in 1930, fourteen years after he died. This may have been a fluke; astronomers today think that Pluto is too small to account for the effects on the orbits of the outer planets, and that there may be a tenth planet that is really responsible for these wobbles. Even so, Lowell's achievements were real and many, and were recognized by the scientific community as much more than the work of a rich dilettante with an interest in astronomy. In 1902, indeed, Lowell was appointed non-resident Professor of Astronomy at MIT, a post he held for the rest of his life, and which involved several series of lectures at the Institute. But perhaps Lowell's greatest contribution to astronomy was to hire a young observer called Vesto Slipher and to set him the task of taking spectra of spiral nebulae and looking for Doppler shifts in their light. Lowell's motivation was his interest in planets – he thought, like many astronomers of the time, that the nebulae might be planetary systems in the process of formation. But the motive for the survey doesn't matter; the results do.

Slipher was in many ways the antithesis of Lowell. Where Lowell was a flamboyant extrovert eager to jump to conclusions, Slipher was quiet and methodical, painstaking, and never willing to announce his discoveries until he had dotted the *i*'s and crossed the *t*'s. The difference in their characters was so marked that it has been suggested that Lowell, aware of his own strengths and weaknesses, deliberately picked Slipher to join the team at the Lowell Observatory to provide the necessary ballast to Lowell's own impulsiveness.

Slipher was born in 1875, in Mulberry, Indiana. He attended Indiana University, graduated in 1901, and was promptly invited to join the Lowell Observatory by Lowell himself. He stayed there for the rest of his career, during which he gained an MA in 1903 and PhD in 1909, both from Indiana University, became Acting Director in 1916 following Lowell's death and then Director of the Lowell Observatory in 1926. It was Slipher who initiated the search which led to the discovery of Pluto in 1930, and although he retired in 1952 he lived until 1969, spanning the time from the days when astronomers still thought that the Milky Way was the entire Universe through to the discovery of radio galaxies, quasars, and the microwave background radiation that is thought to be the echo of the Big Bang itself. The theory that predicted the existence of that background radiation drew in large measure on the line of research begun by Slipher with his measurement, in 1912, of the Doppler shift in the light from the Andromeda Nebula, which we now know to be the nearest large galaxy to the Milky Way.

The Doppler shift, remember, is a displacement of the bright or dark lines seen in the spectrum of light from a moving object. If the object is moving towards us, the shift is towards the blue end of the optical spectrum, and is called a blueshift. If the object is moving away from us there is a corresponding redshift. And the size of the shift, compared with the position in the spectrum of the equivalent lines in light from a stationary object, gives a direct measurement of the speed with which the object is moving towards or away from us, its Doppler velocity.

It's worth pointing out the technical achievement involved in these first measurements of Doppler velocities for nebulae. The 24-inch telescope was a fine instrument, one of the best of its day. But that day just preceded the great leap forward in telescope technology initiated by Hale. Using the best technology available – the best spectroscopes and the best photographic plates – Slipher still had to expose one photograph for twenty, thirty or even forty hours (spread over several nights, of course) in order to obtain one spectrum from which the Doppler shift could be measured. All of this, remember, while working in an unheated telescope dome, because warm air would produce convection currents that would blur the image in the telescope, or in the slit of the spectrograph; working in the cold air at altitude; and keeping, literally, a close eye on the image in the telescope to make sure that the nebula being photographed stayed precisely in position at the centre of the field of view. When he had the photographs, Slipher's problems were far from over. The light from a star is concentrated in a point in the image formed by a telescope, and even when this point of light is spread out by the spectroscope, the spread-out image is still bright enough for the lines to be identified, and for their displacements, if any, to be measured with relative ease. The faint image of a nebula – a galaxy – on the other hand is spread out to start with, and the spectroscopic spreading out of the already faint image makes it fainter still, with the lines in the spectrum difficult to pick out and identify. If the spread-out image is too big, the lines are too faint to be seen; if the image is bright enough for the lines to show up, the chances are it is too small for the shift in the lines to be measured. The success of the Doppler measurement technique depends crucially on the efficiency, or speed, of the photographic emulsions used to record the images.[1] In spite of all these difficulties, in 1912 Slipher obtained four spectrograms of the Andromeda Nebula, M31, which all showed the same clear evidence of a Doppler shift corresponding to a velocity of 300 kilometres per second. This blueshift in the light from Andromeda showed not only that the nebula is approaching us, but that it is doing so at a speed greater than the velocity of any other astronomical object – star, planet or whatever – that had been measured at that time.

Once the breakthrough was achieved, other Doppler velocities for several nebulae were soon measured, although not exactly in a head-long rush. Painstakingly, pushing his equipment to the limit, Slipher extended the list of measured Doppler shifts for nebulae up to thirteen by 1914. Now a pattern began to emerge. Only two of these thirteen measurements showed a blueshift; the other eleven were all redshifts, indicating that the nebulae being studied were all rushing *away* from us, with velocities of hundreds of kilometres a second. It could still have been a coincidence that a preponderance of redshifts showed up in the first spectra measured by Slipher. After all, although you expect a perfectly balanced coin to come up heads or tails in roughly equal numbers when tossed repeatedly, it wouldn't be too astonishing to get a run of thirteen tosses that included only two heads. But as more and more redshifts were measured, the pattern stayed the same. By 1925, Slipher had measured forty-one nebular Doppler shifts, and other astronomers had added four more to the list (some indication of Slipher's achievement – he had measured ten times as many as everyone else put together); now, it was forty-three out of forty-five that showed a redshift, and the record recession velocity measured was up above 1,000 kilometres per second. It began to look like much more than a coincidence, even though at that time astronomers still had no final proof that the nebulae were external galaxies in their own right, far beyond the Milky Way. But they had their suspicions. Arthur Eddington, a great British astronomer who was also a great popularizer of science, wrote in 1923:

One of the most perplexing problems of cosmology is the great speed of the spiral nebulae. Their radial velocities average about 600 kilometres per second and there is a great preponderance of velocities of recession from the Solar System. It is usually supposed that these are the most remote objects known (though this view is opposed by some authorities), so that here if anywhere we might look for effects due to general properties of the world.

By 'world', of course, Eddington meant what we now call the Universe. He went on:

The great preponderance of positive (receding) velocities is very striking; but the lack of observations of southern nebulae is unfortunate, and forbids a final conclusion.[2]

Over the next couple of years an important correction to the Doppler measurements was made when it was confirmed that the Milky Way as a whole is rotating, and a variety of techniques made it possible to estimate the speed of the Sun in its orbit around the centre of the Milky Way. This showed that the Solar System is moving at about 250 kilometres per second in, it happens, more or less the direction of the Andromeda Nebula. That impressive blueshift of 300 kilometres per second is actually mainly due to our motion around the Milky Way; only about 50 kilometres per second is genuinely due to the motion of the Andromeda Nebula towards the Milky Way as a whole. While measured redshifts were setting ever higher speed records, even the couple of known blueshifts were being relegated to much more modest proportions than first impressions had indicated. The stage was set for the work by Hubble and his colleague Milton Humason which gave us the first version of the modern picture of the Universe.

Redshifts rule the roost

In the mid-1920s, there was a suspicion among some astronomers, most notably the German Carl Wirtz, that the largest velocities of recession that Slipher had measured belonged to the most distant of the nebulae being investigated. But this was no more than a suspicion, because nobody, before Hubble, had much idea of the distances to the nebulae. So, naturally enough, it fell to Hubble to put the two pieces of evidence – redshifts and distances – alongside one another and to come up with a redshift–distance relation. After the Sun's velocity around the Milky Way was determined, partly thanks to Wirtz's work, in 1927, the pattern could be discerned even among the forty-odd galaxies whose redshifts had been measured, and that provided the impetus for a major new project to determine redshifts of ever fainter and more distant galaxies, a project largely carried out by Hubble's colleague Milton Humason.

Wirtz had shown in 1924 that the apparent diameters of nebulae as viewed from Earth seemed to be correlated with their recession velocities. He had data for forty-two galaxies, and found that the smaller a galaxy looked the bigger its redshift was likely to be. Assuming all galaxies are really more or less the same size, this immediately suggested that the smaller nebulae only *look* small, because they are more distant, and that therefore greater distance from us and greater recession velocity went hand in hand. But this was only a rule of thumb, since at that time there was no direct measure of the absolute distances to the nebulae. The matter rested until 1929. Slipher had turned his attention to other problems, and only forty-six redshifts of galaxies beyond the Milky Way were known even then. But by then it was clear that these objects *were* galaxies in their own right, and Hubble's development of the pioneering work of Leavitt and Shapley had given him a good idea of their relative distances. Even though we now know that Hubble's baseline calibration was wrong, he could still say, with complete accuracy, that one galaxy was twice as far from us as another, or 1½ times as far, or whatever the ratio might be. And that was all the information he needed. Even so, it is astonishing that Hubble drew the (correct) far-reaching conclusion from a very small number of data.

Hubble had forty-six redshifts, largely inherited from Slipher, but out of those forty-six objects he had distances for only eighteen isolated galaxies and the Virgo Cluster. One obvious way to compare the redshifts and distances of these nineteen objects was to plot a graph of velocity (redshift) against distance. Each object has a unique velocity and a unique distance, so it corresponds to a point on such a graph. When all nineteen points were plotted, Hubble concluded that they lay on a straight line, which meant that velocity must be directly proportional to distance – a galaxy twice as far from us as another galaxy is seen to be receding twice as fast as the nearer galaxy. In truth, this seems a sweeping generalization to make on the basis of the few scattered points Hubble had plotted on his graph (see Fig. 4.1A). You need a lot of faith, and not a little imagination, to draw a straight line through those points and say that the points fall on the line. But astronomers are used to making hazards of this kind, and, as we shall see, it may well be that Hubble already had an inkling of the kind of relationship he was looking for. Shaky though the foundations of the relationship may seem today, further research soon established its basic truth beyond all doubt, and today it is known as Hubble's Law: redshift is proportional to distance. If the law really is universal, and assuming the constant of proportionality (now known as Hubble's constant) has been correctly calibrated, it gives astronomers the ultimate yardstick of the Universe. All they have to do is measure redshift and they know distance. The significance of the discovery goes far beyond that, however, but before we look at the implications we should pay tribute to the man who picked up redshift studies where Slipher left off, and who did more than even Hubble himself to establish the validity of 'Hubble's Law'.

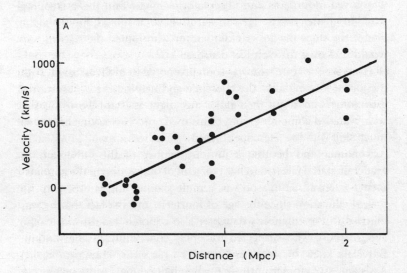

Figure 4.1A One of Edwin Hubble's earliest redshift-distance plots, with just 33 data points and a rather optimistic straight line drawn through them in 1929.

Milton Humason was born at Dodge Center, Minnesota, in 1891. When he was fourteen, he was sent to summer camp on Mount Wilson, and enjoyed himself so much that within a few days of going back to high school he had persuaded his parents to let him take a year off school to go back to the mountain. He never returned to formal education, but by a circuitous route became one of the foremost observational astronomers of his generation. As an academic dropout, he became for a time a mule driver guiding packtrains up the trail to the mountain top while the Mount Wilson Observatory was being constructed. He was fascinated by both the mountain and the observatory work, but also found time to fall in love and marry the daughter of the Observatory's engineer in 1911. With his new responsibility as a married man, Humason gave up mule skinning and tried to settle down on a ranch owned by a relative in La Verne. But in 1917, when a janitor's job at the Mount Wilson Observatory fell vacant, Humason's father-in-law urged him to apply and hinted that a bright young man who loved mountains and observatories might find this a stepping stone to greater things. He can hardly have imagined, however, the size of the steps the 26-year-old janitor who joined the staff in 1917 would take over the next few decades.

From janitor Humason was soon promoted to night assistant, with the job of looking after the telescope and helping the observational astronomers go about their tasks; any night assistant worth his salt soon wangled some observing of his own, and Humason showed so much skill with the telescopes that in 1919 he was appointed Assistant Astronomer and became a junior member of the Observatory's academic staff. Hale had to fight off a lot of opposition to the appointment – after all, Humason was a mule skinner and janitor with no formal education since the age of fourteen, and the fact that he was married to the engineer's daughter also counted against him in the eyes of those who suspected foul play concerning his promotion. But Hale knew his man, and stuck to his guns. Humason stayed Assistant Astronomer until 1954, when he became a full Astronomer at the Mount Wilson and Palomar Observatories; since 1947, he had been Secretary of the Observatories, responsible for public relations and various administrative duties. He received honorary degrees, but never one of the more common kind, and lived to within a few weeks of his eighty-first birthday in 1972. His meticulous handling of delicate instruments and skill with the large telescopes enabled him to provide the base-line data with which cosmologists were able to build their first detailed imaginative models of the Universe, and

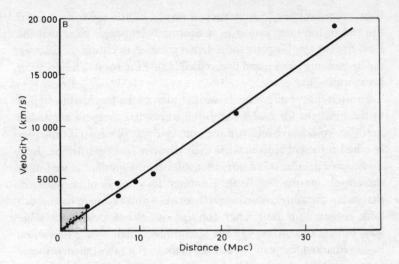

Figure 4.1B By 1931 the accuracy of Hubble's bold mixture of guesswork and science had been confirmed. Hubble and Milton Humason had pushed their redshift studies much further, so that the data of 4.1A all lie within the box on the bottom left-hand corner of this plot, made just two years later. But the straight line is still there, and looks much more believable.

to cast their ideas back to the Big Bang itself. And this all began in 1928, when Hubble first steered Humason towards the task of measuring the redshifts of faint and distant galaxies.

These observations required new instruments, new photographic techniques, and the almost unique combination of patience and skill which enabled Humason to spend hours at the telescope, spread over several nights, guiding it precisely so that he could obtain the spectrum of a far distant galaxy on a tiny photographic plate just half an inch wide. By 1935, he had added 150 redshifts to Slipher's list, and was clocking up redshifts corresponding to recession velocities in excess of 40,000 km per second – more than one eighth of the speed of light. With the advent of the 200-inch telescope, he extended the search still deeper into space, and by the late 1950s the speed record was more than 100,000 kilometres per second, one third of the speed of light, corresponding to a distance of several billion light years. From the end of the 1920s onwards, it became increasingly clear that the Universe is a very big place indeed, that galaxies, or clusters of galaxies, are its building blocks, and that redshifts rule the roost. The Universe is expanding like crazy.

So what does it all mean? It was the turn of the theorists to step in to the limelight for a while, with the somewhat sheepish admission that they could have told you the Universe was like that, if only they had had sufficient faith in what their theories had been telling them for the past decade. It is a story that will become familiar as we follow the search for the Big Bang. Contrary to what is often believed, theorists – certainly cosmological theorists – don't seem to have much faith in their own ideas. They don't go out to bat for them in a big way, but tuck them away in the scientific journals where they often lie unremarked for years.[3] From the 1930s to the 1980s, theorists were repeatedly being surprised by new observational discoveries which, it turned out, were exactly in line with what someone had predicted, half-heartedly, or ignored altogether, ten or twenty years before. And who was the first of these half-hearted theoretical cosmologists, who didn't believe what his own theory was telling him? None other than Albert Einstein.

Einstein

Einstein has become an almost mythical figure, part of the folklore of our times. He is the archetypal genius, the white-haired, slightly eccentric but amiable old man who pierced to the heart of complex problems by applying an almost childlike naïvety and asking questions so obvious that nobody else had thought to ask them. Much of this is true, just as it is true that he was not thought to be particularly bright at school, made no pronounced impact on the academic world as a student, and had to work as a technical expert in a patent office in the early years of this century while he developed three major new ideas in physics in his spare time. But in one respect the stereotypical image doesn't tell us the truth about the man who made these revolutionary contributions to science. In the early 1900s, Einstein was *not* a white-haired, genial patriarch who dressed for comfort rather than elegance and sometimes didn't bother to wear socks. As pictures from that period show, he was a dark-haired, handsome young man who dressed with conventional smartness. This is important, for Einstein's greatest ideas were *youthful* ideas. They provided new insights, overturned established wisdoms, and were truly revolutionary. The burst of activity which brought Einstein to the wide attention of the scientific community was completed in 1905, when he was twenty-six; his greatest achievement, the General Theory of Relativity, was published just ten years later. And although he lived until 1955, becoming the genial old professor of folklore, his greatest works were all behind him before the end of the First World War. Science, especially the mathematical side of science, is like that. Only young minds can stretch to discover and embrace new concepts – and if the new concepts are as dramatically different from old concepts as those Einstein developed, it can take the rest of your life, or several lifetimes, to work out the implications.

Einstein graduated, with no great distinction, from the Swiss Federal Polytechnic in 1900. He had by then already had a chequered academic career – he didn't talk until he was three, and although he got some good school reports and some bad he ended up being expelled from high school (the Munich Gymnasium) at the age of fifteen as a 'disruptive influence'. The expulsion may have been engineered by Einstein – his parents, following a business failure, had already left Germany to live in Italy, and young Albert had such a deep loathing of the militaristic nature of German society that he renounced his citizenship at about this time, becoming stateless rather than be a German. After a happy year of freedom with his family in Milan, in 1895 Einstein applied for admission to the Zurich Polytechnic – and failed the entrance examination. But a year's cramming at a Swiss school in Aarau saw him safely past this hurdle at the second attempt, and he entered the Poly in the autumn of 1896, aged 17½.

It was during this year at Aarau that Einstein began to puzzle over a question that would lead him, ten years later, to the Special Theory of Relativity. Albert puzzled over how a light wave would look if you could run fast enough to catch up with it. Here, indeed, is the prime example of the naïve, childlike approach to the Universe that was to be his hallmark. It is a ridiculous question, isn't it? Like the three-year-old who asks you, 'Why is grass green?' But hold on to it for a while, for there is more to this question than meets the eye.

It was more through boredom than lack of ability that Einstein only scraped through his examinations at the Poly – he didn't bother to study things he wasn't interested in. His arrogance as a student upset his teachers, one of whom, Heinrich Weber, is reported to have burst out 'You're a clever fellow! But you have one fault. You won't let anyone tell you a thing. You won't let anyone tell you a thing.'[4] And it was as much through this alienation of people in a position to help him that Einstein found himself unable to get an academic post, scraping a living as a tutor until he got the famous job at the patent office in Berne in 1902, and became a Swiss citizen. It was a steady job, which he found easy and which gave him both security and plenty of time to think about such puzzles as the nature of a speeding light wave. But nobody could have anticipated the burst of ideas that would come out of his modest hothouse within three years.

Einstein's theories

Volume 17 of the German journal *Annalen der Physik* was published in 1905, and is now a collector's item marvelled over by scientists. In that one volume of the journal, young Einstein, the unknown patent office clerk (he didn't even have a PhD at the time), published three papers giving key new insights into the nature of the world. One of these helped to establish the reality of atoms; another hinted that light might not be simply a wave, but could also behave like a series of particles. Both those papers were to prove important in the development of quantum physics, and it was for the second of them, on the photoelectric effect, that Einstein received the Nobel Prize seventeen years later.[5] But it is the third paper that has become the most famous. It fills thirty pages and carries the unimpressive title 'On the Electrodynamics of Moving Bodies'. That is the paper that tells us that neither space nor time is absolute, but can be squeezed or stretched depending on your point of view; that moving bodies get heavier; that $E = mc^2$; and that points the way to atomic bombs and nuclear power plants, as well as an understanding of what keeps the Sun and stars hot inside. If that sounds impressive, remember that in 1905 what was to prove by far Einstein's greatest work still lay ten years in the future.

The 1905 paper, the foundation stone of the Special Theory of Relativity, is important enough in today's world, but is a minor sideshow on the road to the Big Bang. Einstein's puzzling over the nature of light had its roots in the work of the great nineteenth-century Scottish physicist James Clerk Maxwell, who had set up the equations that describe light as electromagnetic waves moving at a certain speed, commonly denoted by c. The question of what these waves would look like if you could run alongside them at speed c actually provided an important insight into a contradiction between the behaviour of light and the 'common sense' rules we learn from experience in the everyday world. If you ran as Einstein imagined, then the electromagnetic wave would still, presumably, be waving, but as far as you were concerned it wouldn't be moving, in contradiction with Maxwell's equations. Something must be wrong with this view of the world. Maxwell's equations didn't square up with preconceptions based on everyday common sense, and something had to give. Einstein's genius lay in accepting Maxwell's equations, but throwing out those preconceptions to come up with a new, and better, description of reality.

By the early 1900s, experiments had shown that every measurement of the speed of light always gave the same answer, c. Historians of science still argue about whether or not Einstein was aware of these experiments at the time, but that doesn't matter. By subtle arrangements of light beams and mirrors, it is possible to measure the speed of a beam of light moving in the same direction as the Earth through space, or in the opposite direction. Common sense tells you that the answers ought to be different. If I see a bus moving off at 10 miles per hour and run after it at 9 miles per hour in a vain bid to catch it, the speed of the bus relative to me is just 1 mile per hour; if I ride in a bus travelling at 30 miles per hour and another bus on the opposite side of the highway passes in the opposite direction at 30 miles per hour, relative to me the second bus is moving at 60 miles per hour. But light isn't like that. The Earth moves through space at some velocity, which we might as well call v. A light beam overtaking us at velocity c does *not* have a speed $c - v$, and nor does a beam of light approaching us from the opposite direction have a speed $c + v$. Whatever our velocity, and whichever direction the beam of light is coming from, when we measure its speed we always get the answer c.[6]

So, Einstein said, we have to reject our everyday preconceptions. When we are dealing with velocities, one plus one doesn't have to equal two. He worked out a mathematical framework within which the speed of light could always be the same for any observer who measured it from a reference frame moving in a straight line at a steady speed. All of these reference frames can be moving relative to one another (that's where the 'relativity' comes in), but they mustn't be rotating or accelerating (hence the 'special', to show that the theory only deals with certain problems in physics). Everybody in such a reference frame finds the same laws of physics, and is entitled to regard the frame they live in as 'at rest'. And everybody measures the speed of light as c. There is no special reference frame in the Universe.[7]

Without going into details, the results of Einstein's calculations can be summed up simply. The improved law for adding up two velocities v_1 and v_2 is not $V = (v_1 + v_2)$, but rather $V = (v_1 + v_2)$ *divided by* $(1 + v_1 v_2/c^2)$, where c is the speed of light. Because c is so big, 300,000 kilometres a second, for everyday velocities like 10 miles per hour and 30 miles per hour the number you divide by is indistinguishable from 1 – the $v_1 v_2/c^2$ bit is virtually zero. But if you make one of the velocities, v_1 or v_2 (or even both of them) equal to c, strange things start to happen. You can *never* add up two velocities that are less than the speed of light and get an answer that is bigger than the speed of light.

Similar equations fall out of the mathematics to tell us that a moving object gets heavier as its velocity, in our chosen frame of reference, approaches the speed of light, and that at the same time the moving object contracts in the direction of its motion. A moving clock runs slow, compared with one that is stationary in our frame of reference. And, the icing on the cake, the notion of two events occurring simultaneously only has meaning in one frame of reference – an observer moving past you at constant velocity will have a different view of which events precede others or occur at the same time as each other. And all of this is part of engineering today. Machines which accelerate particles such as protons and electrons to close to the speed of light are built in accordance with Einstein's equations. They wouldn't work if the equations were not a good description of the way the world works, and as they work they provide physicists with direct measurements of the mass increase, the time dilation, and other effects predicted by Einstein. Special Relativity works perfectly as a description of the everyday world, marrying up the older mechanics of Newton (still perfectly adequate if you don't deal with things moving at close to the speed of light) and the equations of electromagnetism developed by Maxwell. But it was still only the 'Special' Theory of Relativity. It didn't deal properly with gravity, and gravity is the force that dominates the Universe at large. So it couldn't provide a complete description of the Universe at large. To do that, Einstein needed a more general theory.

The Special Theory was a child of its time. There was an obvious need to reconcile Newton's and Maxwell's ideas, and if Einstein hadn't come up with Special Relativity in 1905 someone else would have, probably within a year or two. But the General Theory was something else again. Nobody except Einstein bothered much about the limitations of the special theory. But after ten years more work (not single-minded work on this one puzzle; Einstein made other significant contributions to quantum theory in those same ten years) he produced a theory that was far more complete than the current *observations* of the Universe. When the observers still hadn't established the distance scale to the nebulae, and didn't know for sure that the nebulae were other galaxies, let alone that those galaxies were almost all rapidly receding from us, Einstein produced a theory which naturally and almost of its own accord described a universe which was a large, empty place which ought to be expanding. Now Einstein had not set out to describe *the* Universe with his equations. He was primarily interested in getting a model of the Universe – a mathematical model – so that he could check that General Relativity could indeed deal with complete universes, and did not run into problems with the conditions at infinity, or at the 'edge' of the universe – the so-called boundary conditions. That was a far deeper mathematical truth. So he was not greatly bothered that his simplest complete and self-consistent theory, which needed no special boundary conditions, did not seem to describe the actual Universe. He was far more interested in the fact that the model was indeed complete and needed no special boundary conditions. In a sense, he didn't accept what the theory was trying to tell him. And for once in his life he did not follow the rule about casting aside all preconceptions. In order to make his model fit more closely to his preconceived idea of the Universe as a static place, he adjusted the equations a little, making them slightly more complicated, to produce a slightly different complete model with no special boundary conditions.

General Relativity is, above all else, a theory of gravity. Almost exactly halfway between the arrival of the Special Theory and the full publication of the General Theory, Einstein published another paper in the *Annalen der Physik*, in 1911, which shows how his mind was working towards a theory of gravity. It was titled 'The Influence of Gravitation upon the Propagation of Light', and although it contains a mixture of half-truths and conjecture rather than any blinding new insight, it points the way forward and brings out another of those naïve questions that reveal deep truths about the Universe. Einstein was struck by the way in which the force of gravity is cancelled out for a falling object – not just for one falling object, but for all of them, in exactly the same way. Galileo had pointed out that all objects fall at the same rate, regardless of how much they weigh; Newton had made use of this insight in formulating his own laws of motion. The effect of force on an object is to produce an acceleration, proportional to its mass – one of Newton's famous three laws. And the size of the force of gravity on an object is also proportional to its mass. So mass cancels out, and all objects fall at the same rate.

Einstein, when he wasn't thinking about the man running after a light ray, seems to have spent a lot of time thinking about another man (or perhaps it was the same one?) trapped inside a falling elevator whose cable had snapped. This was Einstein's way of marvelling at the way things behave when they are falling freely under the influence of gravity. Inside the falling elevator, everything falls at the same rate, and there is no relative motion. The man inside the elevator will float, completely weightless, able to push himself from wall to wall or floor to ceiling with effortless ease. Of course, we have all seen pictures of astronauts doing exactly this inside spacecraft; they can do it for the same reason, they are in 'free fall' under the influence of gravity. Even an orbit around the Earth is a special kind of controlled falling. But Einstein had to *imagine* all the things we have seen for ourselves on TV. A pencil, left weightless in mid-air in the falling elevator; liquids that refuse to pour but form round globules; and so on. The objects inside the falling elevator (or a spaceship) obey the Newtonian laws of motion we learned in school – they move in straight lines at constant velocity unless they are acted upon by some force. In the world outside the falling elevator, things are different because of the force of gravity. Einstein's genius saw the important point missed by everybody else. If the acceleration of the falling elevator can *precisely* cancel out the force of gravity, as it does, that must mean that force and acceleration are exactly equivalent to one another.

Why is this such an important insight? Suppose that the elevator is now replaced in the imagination by a large physics lab, with no windows in it. The lab sits on the surface of the Earth, and a physicist inside can measure how things fall and work out the force of gravity. Now imagine the lab floating in space. The physicist has no trouble working out that he is in free fall. But what happens if the lab is pushed by a steady force, exactly as strong as the force of gravity on the Earth's surface, but in an 'upward' direction, in terms of the arrangement of the floor and ceiling of the lab? Everything in the lab falls to the floor, just as airplane passengers are pressed back in their seats when the plane accelerates on take-off. That pressure soon eases off as the plane levels off at a steady speed. But in our imaginary lab the downward force persists as long as the lab is being accelerated upwards. The physicist can repeat all his experiments, and get exactly the same results that he did when the lab was stationary on the ground. There is *no way* to tell whether the lab is stationary in a gravitational field, or being accelerated upwards. Gravity and acceleration are equivalent.

What has this got to do with light? Go back to the lab being pushed through space by a constant force.[8] The physicist inside might decide to do some experiments involving light. He sets up a light beam so that it starts out on one side of the lab and crosses to the other side. Now, it takes a definite amount of time for the light to cross the lab, and during that time the lab has kept on accelerating upwards, so the wall will have moved up a bit before the light beam reaches it.[9] The physicist can, in principle, measure how far down the wall the spot of light falls, and deduce that his lab is being accelerated. He can even measure the acceleration by measuring how much the beam of light has bent. It looks as if there is, after all, a way to distinguish gravity from acceleration. Not a bit of it, says Einstein. We must keep the idea that gravity and acceleration are equivalent until (and unless) they are *proved* not to be. If the light beam is bent in an accelerating frame of reference, then if the theory is correct it must also be bent by gravity, and by the exactly equivalent amount.

This principle of equivalence is the heart of the correct insight in the 1911 paper.[10] Unfortunately, the calculation of the size of the bending effect was wrong, but no matter. Over the next four years Einstein developed these ideas into a complete general theory of relativity, and the complete theory also predicted that light can be deflected by gravity, indeed by rather more than the amount of deflection calculated in 1911. The best way to understand how this light bending occurs is to cast aside our preconceived ideas about force and space, and to take on board the ideas presented by Einstein, initially in 1915 and in a complete form in 1916. These ideas envisage what we think of in everyday terms as empty space as something almost tangible, a continuum in four dimensions (three of space and one of time) which can be bent and distorted by the presence of material objects. It is those bends and distortions which provide the 'force' of gravity.

Forget about the four dimensions of spacetime for a moment, and think of a two-dimensional elastic surface. Imagine a rubber sheet stretched tightly across a frame to make a flat surface. That is a 'model' of Einstein's version of empty space. Now imagine dumping a heavy bowling ball in the middle of the sheet. It bends. That is Einstein's 'model' of the way space distorts near a large lump of matter. When you roll marbles across the flat rubber sheet, they travel in straight lines. But when the sheet is distorted by the bowling ball, any marble you roll near the ball follows a curved trajectory around the depression in the rubber sheet. That, said Einstein in effect, is where the 'force' of gravity comes from. There isn't really any force. Objects are simply following a path of least resistance, the equivalent of a straight line,

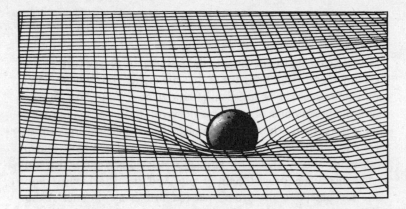

Figure 4.2 The way in which the mass of a body distorts spacetime in its vicinity can be represented by the distortions produced in a smooth rubber sheet by the weight of a heavy ball.

through a curved portion of space, or spacetime. The object can be a marble, a planet or a beam of light. The effect is the same. When it moves near a large mass – through a gravitational field of force, on the old picture – it gets bent. General Relativity predicted exactly how much a beam of light should get bent when it passes near the Sun. The mathematics may be esoteric and the concepts, such as bent space, bizarre. But Einstein's General Theory made a clearcut and testable prediction. It appeared in 1916, when Einstein was working in Germany. The British astronomer Arthur Eddington learned of the new theory, and its prediction, from a colleague in neutral Holland. And this German prediction was confirmed by a British observation made in 1919, when the two countries were still technically at war, having signed an armistice but not yet a peace treaty. Partly for these reasons, it caught the popular imagination like no other discovery in the physical sciences, causing a stir comparable only to the stir caused by the publication of Darwin's ideas on evolution in the previous century.

The proof

A scientific theory can never, strictly speaking, be *proved* correct. The best any theorist can hope for is that his or her theory will make a prediction which can be tested and found to be accurate, to within the limits of observational or experimental error. In that sense, Einstein's theory has proved to be a more complete theory than Newton's theory of gravity, producing predictions which are more closely in agreement with observations. This is the special, restricted sense in which Einstein's theory was 'proved' right in 1919. And the man chiefly responsible for obtaining the proof was the British astronomer Arthur Eddington.

Eddington was three years younger than Einstein, having been born in 1882 in Kendal, Cumbria – the home of the famous 'mint cake' carried by mountaineers up Everest as part of their iron rations. But

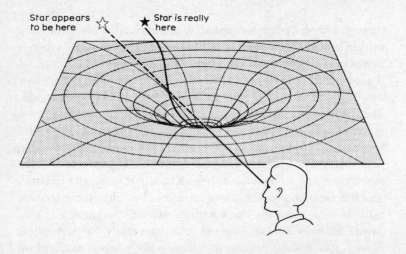

Figure 4.3　Light travelling through spacetime follows the distortions produced by massive objects. Using the image of the ball resting on a rubber sheet, we can represent the distortion of spacetime produced by the Sun, and the effect this has on the light from a distant star.

his father died in 1884, and the young Eddington moved with his mother and sister to Weston-super-Mare, in Somerset, where he was brought up and attended the local school. He was a Quaker throughout his life – something that was to be important to the confirmation of the Einsteinian prediction of light bending, in a roundabout sort of way – and an outstanding scholar who went first to Owens College in Manchester (the college which became the University of Manchester) and then, on graduating in 1902, to Cambridge. Three years later he graduated from the University of Cambridge, and after a short spell teaching in 1907 he became a Fellow of Trinity College, and also took up a post at the Royal Greenwich Observatory as Chief Assistant. In 1912, at the age of 29, he became Plumian Professor of Astronomy and Experimental Philosophy at the University of Cambridge (my favourite academic title) and in 1914 the Director of the Cambridge Observatories.

If all that makes him sound a formidable man, the impression would be only half-correct. Eddington was also a brilliant communicator, who became one of the leading popularizers of science in the 1920s and 1930s, and he had a well-developed sense of humour and of the bizarre. In later life, he told how one of his schoolboy games was to make up sentences which obeyed all the rules of English grammar, but made no sense – one example was 'to stand by the hedge and sound like a turnip'. And in his writings on theories such as quantum physics and relativity he was prone to slip in a bit of Lewis Carroll to help get a point across. There certainly has to be something out of the ordinary run of academics about a man who can begin a chapter of a book titled *Philosophy of Physical Science* with this sentence:

I believe there are 15 747 724 136 275 002 577 605 653 961 181 555 468 044 717 914 527 116 709 366 231 425 076 185 631 031 296 protons in the universe, and the same number of electrons.

Perhaps even more remarkably, the reasons why Eddington came up with this large number are still of interest to cosmologists, as we shall see.

Eddington will be remembered for two great achievements. As much as anyone else, he invented the subject of astrophysics, the study of how physical laws deduced here on Earth, together with observations of the light from stars, can explain the processes going on inside stars which keep them hot, and how the stars must change as they age. And he was also the definitive popularizer of Einstein's theories of relativity in the English language, not just in the sense of communicating these ideas to lay persons, but also as the scientific interpreter who made them clear to his colleagues, and wrote textbooks on the subject which helped to spread its message. Fascinating though all of Eddington's life and work was, the one thing I can pick out here is his response to the prediction that light must be bent when it passes near the Sun.

Einstein's first announcements of the General Theory were communicated to the Berlin Academy of Sciences during the latter half of 1915, and published in more detailed form the following year.[11] Copies of Einstein's papers went, naturally enough, to his friends in the neutral Netherlands, and one of those friends, Willem de Sitter, sent copies of Einstein's papers to Eddington. In 1916 and 1917, de Sitter also sent three of his own papers to the Royal Astronomical Society for publication. These were partly reviews of Einstein's work, explaining its significance, but in the last of the three de Sitter also presented for the first time a description of the Universe, based on General Relativity, which required expansion. More of this in its place. Eddington was Secretary of the Royal Astronomical Society at the time, and we know that he read the papers carefully and reported on them to the Society's meetings, prior to their publication. The one person who had the intellectual ability and background to appreciate fully the significance of Einstein's new work was in exactly the right place, at the right time, to get the news. Fate had several more twists to add to the story before Einstein's new theory was proved correct.

The way to test the light bending, Einstein pointed out, was to look at stars near the Sun during an eclipse. Normally, of course, the bright light of the Sun makes it impossible to see stars in that part of the sky, but with the Sun's light temporarily blotted out by the Moon it would be possible to photograph the positions of stars which lie far beyond the Sun but in the same direction on the sky. By comparing such photographs with photographs of the same part of the sky made six months earlier or later, when the Sun was on the other side of the

Earth, it would be possible to see any shift in the apparent positions of the stars produced by the light-bending effect. What the astronomers needed was an eclipse of the Sun. Ideally, if they could have chosen the eclipse they wanted, they would have asked for one on 29 May, in any year, because just then the Sun is seen passing in front of an exceptionally rich field of bright stars, in the direction of the Hyades. Eclipses are quite frequently visible from some part of the Earth, but an eclipse on 29 May (or any other particular day of the year) is something that happens only very rarely. As Eddington himself commented, 'It might have been necessary to wait some thousands of years for a total eclipse of the sun to happen on the lucky date.' But by a remarkable stroke of good fortune there was an eclipse due in 1919 – on 29 May. It was too good an opportunity to miss, provided that the war was over in time to organize an expedition to observe the eclipse, which would be visible from Brazil and from the island of Príncipe off the west coast of Africa.

In 1917, the plot began to thicken. The Astronomer Royal, Sir Frank Dyson, was enthusiastically in favour of organizing two expeditions to observe the 1919 eclipse, and contingency plans began to be made. Meanwhile, conscription was introduced in Britain, with all able-bodied men eligible for the draft. Eddington was thirty-four and able-bodied; he was also a devout Quaker and a conscientious objector. This was a difficult thing to be in 1917, and the position was further complicated by the recognition of the scientific community that Eddington was a scientist of the first rank. The physics community still felt deeply the loss of Henry Moseley, a pioneering X-ray crystallographer killed in action at Gallipoli in 1915, and questioned the wisdom of the government in sending the best scientists of the day to die, perhaps, in the trenches. A group of eminent scientists pressed the Home Office to give Eddington an exemption, on the grounds that Britain's long-term interests would be best served by keeping him at his proper work. The Home Office eventually agreed, and wrote to Eddington, sending a letter for him to sign and return. But Eddington added a footnote to the letter to the effect that if he were not deferred on the stated grounds he would claim deferment on the grounds of conscience anyway. It was an honest and principled stand, which left the Home Office with a problem, and the scientists who had pleaded on Eddington's behalf more than a little upset. The law of the time said that a conscientious objector must be sent to do not very congenial work in agriculture or industry; Eddington was quite prepared to go and join his Quaker friends. The upshot of a further round of debate, involving Dyson, as the Astronomer Royal, was that Eddington's draft was deferred, but with the 'condition' that if the war ended by May 1919 he *must* lead an expedition to test the light-bending prediction of Einstein's theory![12]

Eddington had led an expedition to Brazil to study the 1912 eclipse of the Sun. And he needed all his experience to ensure the success of his part of the twin 1919 expeditions, which were planned throughout 1918. The idea was that Eddington and a Cambridge team would go to Príncipe, while Dyson would organize a team from the Royal Observatory, Greenwich, to observe the eclipse from Brazil. But no work could be done by the instrument makers until the armistice was signed. They were too busy building weapons of war. But the expeditions had to sail in February 1919. The armistice was signed of course, on 11 November 1918, and the anniversary is still marked by memorial services. In a few hectic weeks, everything was made ready and the expeditions set off. The Brazil expedition had perfect weather for the occasion and obtained a series of excellent photographic plates of the star field around the Sun at the time of eclipse. But for logistical reasons these plates were not processed and studied immediately. On Príncipe, Eddington waited anxiously as the appointed day dawned rainy with a cloud-covered sky. More in hope than expectation, all the arrangements to photograph the eclipse were made, and just near the time of totality the Sun showed dimly and the plates were exposed. The result was just two plates showing the stars needed for the test. Eddington had arranged for these plates to be examined on the spot, 'not entirely from impatience,' as he put it, 'but as a precaution against mishap on the way home.' One of the successful plates was duly developed and analysed on Príncipe, Eddington comparing it with another plate of the same part of the sky that he had brought with him. The measurements required were simple. Three days after the eclipse, Eddington knew that he held in his hand the proof that Einstein's General Theory of Relativity was right.

The full analysis of the eclipse observations took several months,

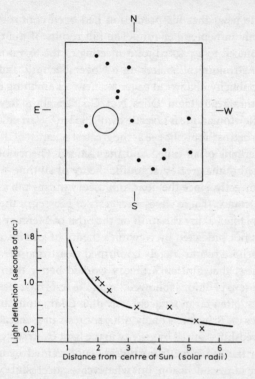

Figure 4.4 During the eclipse of 1919. Arthur Eddington was able to measure the positions of several stars, shown in the top diagram, which lay in nearly the same direction on the sky as the Sun (circle) at the time of the eclipse. The stars are, of course, much further away from us than the Sun is, but this juxtaposition means that light from the distant stars passes through the region of space affected by the Sun's gravity, as in Figure 4.3.

When these positions were compared with the measured positions of the same stars when the Sun was on the opposite side of the sky, Eddington found that they were apparently deflected, each by an amount which depended on the angular separation of the star from the Sun at the eclipse. Light had been 'bent' as it passed by the Sun. And these 'deflections' (crosses in lower figure) fell exactly on the curve predicted by Einstein's theory.

and definite news that his prediction had been confirmed reached Einstein only in September 1919. The full results of the expeditions were announced to a packed joint meeting of the Royal Society and the Royal Astronomical Society on 6 November 1919, and produced a wave of publicity in a world eager for news of anything except war. The headlines read 'Light Does Not Go Straight', 'Revolution in Science', 'Newtonian Ideas Overthrown', 'Space "Warped" '. Einstein was established in the public eye as the greatest scientist of the twentieth century, perhaps of all time. And the General Theory of Relativity was accepted as the greatest scientific theory of all time – more than a little incorrectly, since the quantum theory must rank as of at least equal importance. There were other tests of Einstein's theory. It had already explained a tiny variation on the orbit of Mercury around the Sun that is not predicted by Newton's theory of gravity, and so in a sense the eclipse results merely confirmed to astronomers what they already knew, that Einstein's theory worked better than Newton's. Other eclipse expeditions followed,[13] and the tests have been repeated many times, often far more accurately than Eddington's first analysis of his plates on Príncipe. Totally different tests of General Relativity, involving redshifts caused by gravity in the light from stars, and subtle changes in the radiation from pulsars (undreamed of in 1919), all point to the same conclusion. But whatever its successful explanations prior to 1919, and whatever tests have been carried out since, 29 May 1919 stands as the day when science made the observations that proved Einstein correct, and 6 November 1919 stands as the day the public were made aware of the fact. Meanwhile, though, the astronomers had a puzzle. If Einstein's theory was such a good description of space and time, why did it say such peculiar things about the Universe?

Einstein's Universe

General Relativity is about the geometry of the Universe – the geometry of spacetime. For the past fifty years or more, one of the cosmologists who has made himself most familiar with the equations and meaning of General Relativity is Bill McCrea, still active as Professor Emeritus Sir William McCrea, at the University of Sussex. McCrea was born in 1904, and graduated from Trinity College, Cambridge, in 1926. He was just of the generation which was the first to receive Einstein's ideas fresh, at the start of their academic careers. Early in a long and distinguished career, McCrea was responsible, with Edward Arthur Milne, for showing how even the rules of Newton's theory of gravity, subject to certain simplifying assumptions, make the same kind of predictions about the evolution of the Universe, especially its expansion, that come out of General Relativity. He has also investigated galaxy formation, studied the significance of those large numbers that so fascinated Eddington, and made important contributions to quantum physics and stellar astronomy. There is nobody better around today to turn to for a lucid explanation of what General Relativity is all about, and I learned my own cosmology as a student of McCrea's in the late 1960s.[14] So, without stumbling through all of the historical process which led up to the modern understanding of General Relativity, and with suitable apologies to Einstein and Eddington, I shall give you my own interpretation of General Relativity.

Special relativity combines space and time into one physical identity with a specific mathematical description, spacetime. This spacetime is geometrically 'flat' – mathematically speaking, it has the same kind of geometry as the flat surface of a floor, or of a billiard table. So it is a special case among a family of more general possibilities, a family of curved surfaces. To the mathematician, a 'curved' surface means anything that is not flat – the undulations of mountains and valleys over the surface of the Earth, as well as the way the surface of the Earth is wrapped around to make a more or less spherical globe. A perfectly spherical surface is as much a special case as a perfectly flat floor, and even curved surfaces can have ripples, like the mountains and valleys on the surface of the Earth.

In a similar way, spacetime can be thought of as curved, and this is the generalization that makes 'General Relativity'. When they look at the Universe and try to describe (or predict) its behaviour mathematically, cosmologists are selecting out one or two families of curved surfaces from the array of possibilities to see which ones fit best. The chosen examples are called models, but they have no physical reality like a Plasticine model; they exist only in the minds of the cosmologists and in their equations. In such models, only the broad features are interpreted, and the lesser ripples, comparable with the mountains and valleys on Earth, are on too fine a scale (if they exist at all) to enter into the calculations. But the distortion in spacetime produced by the Sun which causes the bending of starlight as it passes nearby is just such a ripple in the fabric of the Universe.

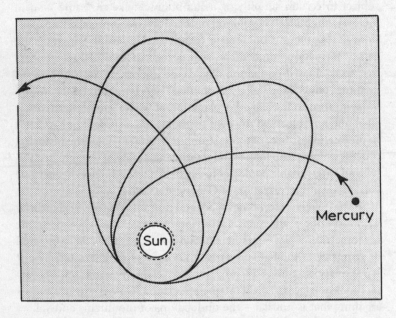

Figure 4.5 The planet Mercury follows an elliptical orbit around the Sun, but the whole orbit also moves around, tracing out a pattern like the petals of a daisy. This movement of the orbit is unexplained by Newton's theory of gravity, but is predicted by General Relativity.

General Relativity goes further than Special Relativity in another way. As I have explained, it describes matter as well as space and time. Special Relativity gave us spacetime; General Relativity in effect gives us 'matterspacetime', although I have never seen the term used as such. It is the matter that bends, or distorts, the fabric of spacetime, and General Relativity gives a well-defined physical meaning to a completely specified geometry of matter, space and time – a universe, where the lower case means that we are referring to just one of the many possible mathematical models, not necessarily to the actual Universe we live in. Interestingly, strictly speaking, General Relativity deals *only* with complete universes. When Einstein's equations are used to describe the way light bends as it passes the Sun, or how the orbit of Mercury shifts slightly each time the planet moves around the Sun, they are being used in an approximate sense. In practice, these approximations can be made as accurate as you like. There are what are called 'boundary conditions' that join on the equations describing a little local object like the Sun to the rest of the Universe. But the important point is that Einstein did not have to expand his theory, in some sense, to make it capable of dealing with the whole Universe. General Relativity, from its birth, dealt with whole universes quite happily; the subtlety is focusing it down to deal with such a minor and insignificant part of spacetime as our Solar System.

In 1917, the received wisdom was that our Milky Way Galaxy was the entire Universe, a stable collection of stars. Individual stars might wander about within the cloud, but taking the Milky Way as a whole, the pre-eminent feature of the Universe seemed to be stability. The Milky Way wasn't getting any bigger or any smaller, and had been there, for all anyone knew, throughout eternity. Some stars might be born and others die, but the overall appearance of the Milky Way stayed much the same, in a steady state. So when Einstein took the wonderful new tool he had invented and applied it to make a description of the Universe, he expected it to at least allow for the possibility of a universe existing in such a steady state. In another paper to the Berlin Academy of Sciences, he described in 1917 his own surprise at what the equations told him, and how he managed to force them to fit into the box of his preconceptions. The paper was called 'Cosmological Considerations on the General Theory of Relativity'; McCrea quotes Einstein (in translation) as saying in the paper, 'I shall conduct the reader over the road that I have travelled, rather a rough and winding road, because otherwise I cannot hope that he will take much interest in the result at the end of the journey.' And McCrea describes the whole paper as 'tentative ... out of character for Einstein.'[15] Truly, the great man was baffled by the fruits of his own labours – because, we now know, the *observations* of the Universe were misleading.

Einstein tried to describe the simplest possible model universe that bore any relation to reality, one which contained matter spread out uniformly through space. The universe was closed, like the surface of a sphere, wrapped around on itself so that there could be no edge.[16] But it wouldn't stay still. When Einstein put the equations describing such a static universe through the appropriate manipulations of General Relativity, they said that the universe must be either expanding or contracting, but that it could not stand still. The only way he could hold the model universe still, to mimic the appearance of the Milky Way, was to add an extra term to the equations of General Relativity, a term called the 'cosmical constant' and often represented by the Greek letter lambda. Then, and only then, would the equations provide a description of a universe that neither expanded nor contracted. The very last sentence of Einstein's 1917 paper reads, 'That term is necessary only for the purpose of making possible a quasi-static distribution of matter, as required by the fact of the small velocities of the stars.' A dozen years later, Hubble had shown that the Milky Way was not the entire Universe, and had discovered the very *large* velocities, all of recession, of the distant galaxies. The reason for the existence of the lambda term had gone, and the existence of the expanding Universe remains the greatest prediction Einstein never made. What Einstein's own equations had been trying to tell him, that we live in a dynamic, expanding Universe, not a static universe, was plain for all astronomers to see. Einstein later described the introduction of the lambda term as 'the biggest blunder of my life'; but it is hard to see how anyone studying the nature of the Universe in 1917 could have failed to make such a blunder. Theory had run ahead of the observations, and it was only when Hubble and Humason showed that we do live in an expanding Universe – Einstein's Universe – that General Relativity at last took its rightful place as a description of the origin and evolution of the entire Universe.

Chapter 5

The Cosmic Egg

Modern cosmology began with Einstein's General Theory of Relativity, and his first cosmological paper, in 1917, in which he struggled to make the equations of relativity fit the mistaken belief of the time that the Universe was static and eternal. But these ideas did not spring out of nothing at all. The seeds, in the form of ideas about curved space and non-Euclidean geometry, were planted by the middle of the nineteenth century. And even the idea that these concepts might have some bearing on the real Universe was voiced before the end of that century, by the Englishman William Clifford, who unfortunately died in 1879 (the year Einstein was born) at the age of thirty-four, before he had time to develop his thoughts into what might have become the first modern cosmological model.

The pioneers of the mathematical study of non-Euclidean geometry included Karl Friedrich Gauss, the brilliant German theorist who lived from 1777 to 1855: although his investigation of geometries in which parallel lines do not behave in the common-sense way of the everyday world and of Euclid's geometry was fairly limited, he deserves pride of place for coining the phrase which translates as 'non-Euclidean geometry'. Pride of place for working out a *comprehensive* non-Euclidean geometry, however, goes jointly to the Russian Nikolai Ivanovich Lobachevski (immortalized in an entertaining ditty by Tom Lehrer) and the Hungarian Janos Bolyai, working independently of each other in the 1820s. Lobachevski and Bolyai each came up with specific kinds of non-Euclidean geometry; in 1854 another German, Bernhard Riemann, put this branch of mathematics on a secure footing by looking at the whole basis of geometry, in a mathematical sense, and laying the groundwork for a whole range of different possible geometries, each as valid as each of the others, with Euclid's geometry included as just one example of the wealth of mathematical possibilities. And Riemann's work, by its generality, allowed for the possibility of developing geometry into the realms of more than three dimensions. The new mathematics that developed in the wake of Riemann's work became a vital tool for Einstein; it is Riemannian geometry that, among other things, allows the possibility of investigating mathematically a theoretical model of the Universe which is a four-dimensional analog of a sphere, a 'hypersphere' in which each of the three dimensions of space that we experience directly is curved, with constant curvature through a fourth dimension, doubling back upon itself in just the same way that 'straight lines' drawn upon the surface of a globe double back upon themselves, to make a 'closed' universe.

Clifford helped to introduce these ideas to the English-speaking world. He translated Riemann's work into English, and like Riemann he was certainly aware of the possibility of a 'finite but unbounded' universe, the higher-dimensional equivalent of the surface of a sphere. It was in 1870 that he read a paper before the Cambridge Philosophical Society in which he talked of 'variation in the curvature of space' and made the analogy that 'small portions of space *are* in fact of nature analogous to little hills on a surface which is on the average flat; namely, that the ordinary laws of geometry are not valid in them.'[1] J. D. North, in his book *The Measure of the Universe* (Oxford University Press, 1965), mentions more than eighty scientific papers on the statics, dynamics and kinematics of non-Euclidean geometry published in the half-century ending in 1915. Any of those authors might, perhaps, have made the leap of imagination needed to suggest that the resulting equations could be applied to the real Universe. But they did not. The required imaginative leap had to await the man who was never afraid to move into uncharted scientific territory, and who built a complete new theory of gravity *and* the Universe with the aid of the tools provided by Riemann and his contemporaries. Einstein started the ball rolling. But it was to be another decade before anyone began to take seriously the idea of applying these equations as a description of the real Universe – and it was to be another half-century before cosmologists realized just how 'real' the resulting description of reality might be.

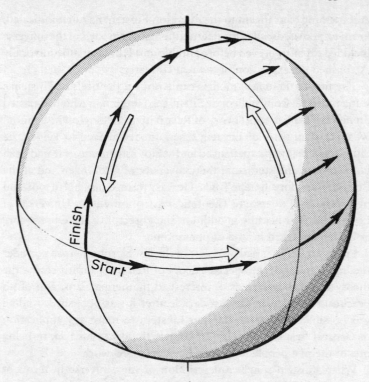

Figure 5.1 A simple example of the oddities of non-Euclidean geometry. On the surface of a sphere, if it is possible to take a little arrow (a vector) on a journey around the equator, up to the pole and back to its starting point. The direction in which the arrow points is very carefully kept the same on each stage of its journey. But when it gets back to the start, the arrow is pointing due north instead of due east!

The fathers of cosmology

Einstein is perhaps best regarded as the 'grandfather' rather than the 'father' of modern cosmology. He pointed the way and set the ball rolling. But it was others who picked the ball up and ran off with it, at something of a tangent to the direction Einstein had first indicated. From 1917 onwards, *all* of the new mathematical models of the universe included expansion, even before Hubble and Humason unequivocally established the expansion of the real Universe.

The first of these new models came hot on the heels of Einstein's static universe, from Willem de Sitter, the Dutchman who had passed on news of the General Theory of Relativity to Eddington in London. By 1915, when news of Einstein's new theory arrived in Leiden, de Sitter was already an experienced and senior astronomer. He had been born in 1872, and studied at the University of Groningen and at the Royal Observatory in Cape Town. He was awarded his PhD in 1901, and by 1908 he was Professor of Theoretical Astronomy at the University of Leiden. He later became in addition the Director of the Observatory at Leiden, and died in 1934 of pneumonia.

Professor de Sitter had been one of the few astronomers to consider the implications of the Special Theory of Relativity for his craft – the theory was mostly seen as of interest to mathematicians, and of no practical relevance, in the first decade after it was published. And he was possibly the first person after Einstein to make any application of General Relativity, in part, of course, thanks to his luck in being one of the first people to hear of Einstein's new work.

When Einstein sought a description of the Universe in terms of General Relativity, that is exactly what he sought – *a* description, a single, unique solution to the equations. His static model, with the cosmological constant, seemed to fit the bill. But in one of the papers de Sitter sent to the Royal Astronomical Society in London in 1917, where it was read with great interest by the then Secretary of the society, Arthur Eddington, he showed that there was another solution to the equations, a solution which represented a different model universe. Obviously, both solutions could not represent the real world. We now know that neither of them represents the real world, and that is not seen as a problem. But at the time this was something of a blow to Einstein's theory, since it could be argued that if the theory offered a choice of universes, all consistent with the basic equations, it couldn't be telling us much, if anything, about the real Universe. That argument scarcely held up once the expansion of the Universe was understood, and astronomers realized that Einstein's equations had indeed been predicting this for at least ten years before Hubble's announcement of the redshift–distance relation.

De Sitter's universe, like Einstein's, was in a mathematical sense static (and, like Einstein's, it included a cosmological constant). Unlike Einstein's, however, it contained no matter at all – it was a mathematical description of a completely empty universe. It is difficult to know what 'static' means in a completely empty universe, since there is nothing that can be used as a marker against which motion can be measured. And when the theorists tried the mathematical equivalent of sprinkling a few specks of matter into de Sitter's universe they found a curious thing – these specks of matter, test particles, promptly rushed away from one another. Furthermore, when they calculated how light from one of the test particles would look as seen from one of the other test particles, the mathematicians found a redshift proportional to the distance between the particles. The de Sitter universe only seemed to be static because it was empty; in a universe containing just a little bit of matter, a few galaxies scattered here and there throughout space, astronomers would see exactly the redshift– distance relation that Hubble and Humason were to find in the late 1920s. Much later, Eddington summed up the distinction between the first two relativistic cosmologies: Einstein's universe contains matter but no motion; de Sitter's universe contains motion but no matter.[2]

Eddington was one of the few people to take the expansion of a de Sitter universe containing a trace of matter seriously at the time. From the limited redshift data that were coming in in the early 1920s, mainly from Slipher, he concluded that de Sitter's variation on the relativistic theme was telling astronomers something about the real world. He later developed his own variation, a model in which the universe sat for a long time (perhaps an infinitely long time) in a static state, like the Einstein universe, and then began to expand, like the de Sitter universe, as galaxies formed. But this model soon turned out to have little bearing on the nature of the real Universe. Einstein and de Sitter themselves had important second thoughts about their cosmological models once the redshift–distance relation began to look like an important feature of the real Universe. In 1932, they put their heads together and came up with yet another model universe, the Einstein– de Sitter model (not to be confused with either of their solo efforts) in which they went back, in a sense, to the roots. The cosmological constant had originally been introduced to hold the model static, but the real Universe was seen to be expanding – so, away with the constant. The earlier models had involved curved space (and, in de Sitter's case, curved time as well), but there was no direct evidence that space was curved – so, away with curved space (but not with curved *spacetime*). The Einstein–de Sitter universe is the simplest universe that can be constructed using the basic equations of General Relativity. It expands, as the equations require, and the space which is expanding is flat, the space of Special Relativity. And because nothing is added to the model to stop the inevitable happening when we look back in time, the model requires that there was a definite creation event, long ago, when the universe was born out of a mathematical point, a state of infinite density, called a singularity.

As the simplest solution to the equations, the Einstein–de Sitter model is a very useful case study. It has pride of place in many courses on cosmology today, on the reasonable grounds that students ought to start with the simplest examples and work their way up to more complex, and more interesting, things. But it is far from clear just how much pride Einstein and de Sitter had in their creation. Chandrasekhar reports Eddington's comments just after the Einstein–de Sitter paper appeared in 1932:

Einstein came to stay with me shortly afterwards, and I took him to task about it. He replied, 'I did not think the paper very important myself, but de Sitter was keen on it.' Just after Einstein had gone, de Sitter wrote to me announcing a visit. He added: 'You will have seen the paper by Einstein and myself. I do not myself consider the result of much importance, but Einstein seemed to think that it was.' (*Eddington*, page 38.)

With hindsight, the key feature of the Einstein–de Sitter model, and the reason it is so widely taught, is that it includes that moment of creation, what we now call the Big Bang. But the hesitant way in which the two pioneers put forward their model perhaps indicates their own discomfort with this idea – an idea which was not theirs, but which had come independently from the two other founders of modern cosmology, in the 1920s, and gained attention only after Hubble's publication of the redshift–distance relation. The Big Bang

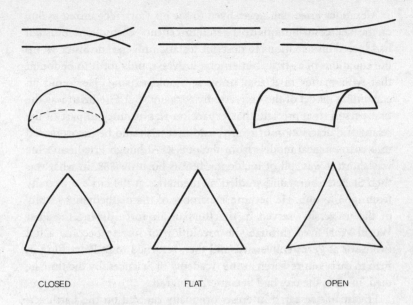

CLOSED   FLAT   OPEN

Figure 5.2 Space may conform to one of three basic geometries. Although our space is three-dimensional, we can see the three possibilities by looking at the curvature of a two-dimensional 'universe'.

If it is positively curved, or closed, like the surface of a sphere, then 'parallel' lines must eventually cross, and the angles of a triangle add up to more than 180 degrees.

If it is negatively curved, or open, like a saddle surface, then parallel lines diverge from one another and the angles of a triangle add up to less than 180 degrees.

Flat space corresponds to a special case, dividing the closed universes from the open universes. Only in this very special universe do parallel lines stay the same distance apart, and the angles of a triangle add up to precisely 180 degrees. Our Universe is very nearly flat. Explaining why this should be so is a fundamental cosmological problem.

was to become fully respectable only in the 1940s; and it wasn't to become the *dominant* theory in cosmology until the 1960s.

Alexander Friedman never lived to see his work recognized as one of the key contributions to twentieth-century cosmology. Einstein had forced his equations to describe a static universe; de Sitter set up the equations of a static, but empty, universe, only for it to be found that with matter in it that universe would expand. He found an expanding model of the Universe by accident. But Friedman was the first person to appreciate that expansion was an integral part of the relativistic description of reality, and that it should be incorporated into cosmological models from the outset. Although Friedman's life was short, it was full of incident. He was born in 1888, in what was then St Petersburg, and studied mathematics at the city's university from 1906 to 1910. He became a member of the mathematics faculty of the university, served in the Russian air force during the First World War, lived through the revolution of 1917 to become a full Professor at Perm University, and then returned to St Petersburg in 1920 to carry out research at the Academy of Sciences. By the time he died, in 1925, the city had become Leningrad.

Friedman's research interests originally centred on the Earth sciences – geomagnetism, hydromechanics, and meteorology. But as an able mathematician he was keenly interested in Einstein's work, and in 1922 he published his solutions to the cosmological equations of General Relativity. There were two key features of this work which remain fundamental to modern cosmology. First, from the outset Friedman realized he was dealing with a *family* of solutions to the equations. He understood that there was no unique solution, as Einstein had hoped, but instead a set of different variations, each describing a different kind of universe. Second, Friedman incorporated expansion into his models as of right. In a way, this echoed the work of Clifford in the 1880s, the idea that space might be uniformly curved, like the spherical surface of a soap bubble, but that this curvature might be changing with time – decreasing, perhaps – as the 'bubble' expands. Friedman's models offered several variations on the theme.[3] In some versions, the bubble expanded forever; in others, it expanded up to a certain limiting size and then collapsed back upon itself, as the force of gravity overcame the expansion. There were versions with a cosmological constant, and alternatives – the preferred alternatives today – in which the cosmological constant was set to be zero.[4] But in all of the models there was at least a period – an interval of time – during which the whole universe expanded in such a way that it would produce a recession velocity proportional to distance.

Friedman also appreciated a point which cannot be stressed too highly, and merits repetition. The redshift in the expanding Universe (or universes) is not caused by the galaxies moving apart from one another *through* space. It is caused by space itself stretching, like a stretching rubber sheet, between the galaxies. Space – or better, spacetime – expands, and carries the galaxies along with it for the ride.

It is something of a mystery why Friedman's work, which was published in a well-known and widely read journal, was ignored. It was drawn to Einstein's attention by one of Friedman's colleagues on a visit to Berlin, and he acknowledged its accuracy in a brief note to Friedman, but even Einstein failed to realize that it might be telling us about the real Universe. Mathematicians in the 1920s had little contact with astronomers; the astronomers seldom followed new developments in mathematics; and Europe and America were much more separate scientific worlds then than they are today. So new mathematical ideas in Europe were not immediately matched up with new astronomical observations being made in the United States. And could there have been an element of prejudice among the theorists against the Russian mathematician better known for his work in meteorology? Whatever the reasons, Friedman died in obscurity in 1925 – and, ironically, it may well have been his interest in meteorology that caused his death. Many of the official biographies report that Friedman died of typhoid. But according to the cosmologist George Gamow, the cause of his death was pneumonia contracted following a chill Friedman caught while flying in a meteorological balloon. Gamow, who became a key figure among the next generation of cosmologists, ought to have known, since he was one of Friedman's students, and only moved to the United States in the middle 1930s.[5]

So it was left to the next person to solve Einstein's equations in the same way as Friedman – but quite independently, and with no knowledge of Friedman's work – to make the breakthrough into seeing these solutions accepted as a worthwhile tool for cosmologists probing the nature of the Universe. The breakthrough was made by

Georges Lemaître, a Belgian cosmologist. His original publication, in an obscure Belgian journal in 1927, attracted little notice. But following the announcement of the redshift–distance relation, the ubiquitous Eddington learned of Lemaître's paper and arranged for an English translation, which appeared in the *Monthly Notices of the Royal Astronomical Society* in 1931.[6] If anyone deserves the title 'father of the Big Bang' it is Lemaître – which has led to some awful puns over the years, since as well as being a cosmologist and mathematician, Lemaître was a priest.

He had been born in 1894 and trained as a civil engineer. During the First World War, he served as an artillery officer with the Belgian Army. After the war, he studied at the University of Louvain, graduating in 1920, before joining a seminary and being ordained as a Roman Catholic priest in 1923. He then spent a year in Cambridge, where he worked with Eddington, and a year in the United States, divided between Harvard and MIT, before returning to Louvain, where he became Professor of Astronomy in 1927 and stayed for the rest of his career. Throughout his long career – he died in 1966 – Lemaître continued to develop his cosmological ideas, and he lived to see many of them incorporated into mainstream cosmology. The most important of these was the idea of the Big Bang itself, although he did not give it that name. Lemaître's equations were essentially the same solutions as those found by Friedman, although he preferred throughout his life to retain the cosmological constant which even Einstein had abandoned in the 1930s. But unlike Friedman, or indeed anyone else before him, Lemaître tackled the question of what those equations were telling about the origin of the Universe.

Unlike Friedman, Lemaître clearly knew something of the observations of galaxy redshifts being made at the time. In his 1927 paper he recognized that galaxies might provide the 'test particles' by which universal expansion could be measured, and he gave, without any reference, a value for the constant of proportionality in the redshift–distance relation (what later became known as Hubble's constant) so close to the value Hubble published a little later that, as one modern cosmologist has put it, 'There must have been communication of some sort between the two.'[7] Lemaître saw what both the observations and the theory of relativity were saying. If galaxies are far apart today, and getting further apart, that must mean that they were closer together in the past. If we look back far enough into the past, there can have been no empty space between the galaxies. Earlier still, there must have been a time when there was no empty space between the stars, and even earlier a time when there was no space between the atoms, or between the nuclei that lie at the hearts of atoms. That was as far back as even Lemaître's bold imagination took him. He envisaged a time when the entire content of the Universe was packed into a sphere only about thirty times bigger than our Sun, what he called a 'primeval atom'. This atom then exploded outwards, breaking up into fragments that became the atoms, stars and galaxies that we know, with the galaxies separating from one another because of the expansion of the universe. The process has been likened to the way in which an unstable, radioactive atomic nucleus may spontaneously split into pieces that go their separate ways – nuclear fission, the power source of the atomic bomb. This simple idea has been developed and modified considerably since the 1930s. But modern cosmology retains at its heart the idea, first propounded by Lemaître, that our Universe was born out of a superdense state which gave birth to everything we can see in the expanding Universe. With the observations by Hubble and Humason which showed our Universe to be expanding, and with Lemaître's idea of the primeval atom in print in English in the early 1930s, modern cosmology was off and running.

Of course, there were other pioneers, even in the 1920s, who made important contributions to the development of relativistic cosmology. The American Howard Robertson, a mathematician who extended de Sitter's work and made a major contribution to the mathematical basis of cosmology, certainly deserves a mention. Together with his English colleague Arthur Walker he developed, in 1935, a mathematical description of homogeneous and isotropic spacetimes called the Robertson–Walker metric, which describes universes with uniformly curved space but a cosmic time that is the same for all observers that move with the expansion of the universe. Such idealized universes, the Robertson–Walker models, are widely discussed today. But the key theoretical developments came from four people, exactly during the ten years or so that Slipher, Hubble and Humason were gathering the evidence that would show the Universe to be expanding.

These 'fathers of cosmology' were Einstein, de Sitter, Friedman and Lemaître – with Eddington, perhaps, as the benevolent godfather who helped the infant along the first faltering steps of the road to maturity. By the early 1930s, theory and observation had come together in a remarkable fashion to point firmly in the direction of the Big Bang. It was another decade before George Gamow, Friedman's former student, developed a fully worked-out version of the new cosmology. Meanwhile, the various alternative models had been developed more fully. Leaving aside some bizarre variations which were espoused at various times by individual cosmologists, the solutions to the equations – solutions unveiled by Friedman and Lemaître in the 1920s – fall into three main categories.

The age of the Universe

One reason why the new cosmological models failed to take the scientific world completely by storm in the 1930s was the problem of the time scale implied by the best interpretation then available of the redshift data gathered by Hubble and Humason. The redshifts give a measure of how quickly galaxies a certain distance apart from one another are separating.[8] If we make the simplest possible assumption, that the Universe has been expanding at the same rate ever since it emerged from the Big Bang, it is very easy to calculate the time that has elapsed since the Big Bang – the 'age of the Universe', if you like. When astronomers did this in the 1930s they hit a snag. The recession velocity is equal to Hubble's constant, H, multiplied by the distance between two galaxies. If the expansion has continued always at the same rate, the time since the beginning of the expansion – the time since the two galaxies were touching – is just $1/H$. Using the value of the Hubble constant deduced by Hubble himself they found that the implied age of the Universe was only about two billion years (2×10^9 years).

This was embarrassing, because there was already good evidence that the Earth and the stars (including our Sun) are older than that. Over the next twenty years or so, the conflict got worse, as a variety of different techniques clearly indicated that most of the things we

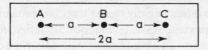

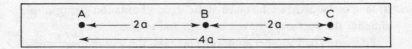

Figure 5.3 Expanding spacetime is like stretching a piece of rubber. The 'galaxies' A, B, and C do not move through the space between them. But when the space expands to double the distance between A and B, it also doubles the distance between every other pair of galaxies, including A and C. From the viewpoint of every galaxy in this universe, every other galaxy is receding at a rate which is proportional to its distance. Because C is twice as far away from A as B is, for example, when all distances are doubled (when the scale factor doubles) it seems that C has 'moved away' from A twice as fast as B has.

can see in the Universe are older than this simple estimate for the 'age of the Universe' itself. Geological evidence, and measurements of radioactivity and the remains of radioactive atoms in samples from the Earth (and more recently the Moon) and from meteorites point to an age of at least four billion years, and probably close to 4.6 billion years, for the Solar System; as an understanding of nuclear physics developed in the 1930s, astronomers (with Eddington prominent among them) began to work out what kept the Sun and stars hot for such a long time, and inferred that many of the stars and galaxies we can see had been around not even just for four billion years but for 10 billion years or more.

One way around the difficulty was to make the cosmological models more complicated, providing a breathing space in which everything could age. After all, there was nothing in the equations which said that the Universe *must* have been expanding at the same steady rate ever since the Big Bang. Lemaître favoured a variation on the theme which started out from the primeval atom, but which included a cosmological constant slightly bigger than the one in Einstein's static model. Such a universe expands fairly quickly at first, then slows down and moves over into an almost static state where it hovers for a time until the expansion picks up again. By choosing the exact value of the constant appropriately, the Lemaître model could be made to stay for as long as you liked in the quasi-static state, a hesitation giving ample time for the stars and galaxies to form and evolve.

Eddington had another solution. He found the idea of the primeval atom and the Big Bang 'unaesthetically abrupt', and pointed out that the Universe could have existed in a static state from the beginning of time (whatever that meant) until some disturbance caused it to start to expand a few billion years ago. Such ideas gained few adherents, apart from their own proponents.[9] They were considered by most astronomers to be artificially forced into an unnatural mould, just as Einstein had originally forced his cosmological equations to fit the static picture of the Universe.[10] And, once again, it turned out that it was the picture of the Universe that astronomers had that was wrong, not the simpler versions of the cosmological equations.

If the Universe really has to be 10 billion years, or more, old and the simple interpretation of the expansion was correct, then it must mean that the accepted value of Hubble's constant was too large. Because the age of the Universe goes as $1/H$, halving the constant, for example, would double the calculated age of the Universe. And in the early 1950s there was just such a dramatic revision of the cosmological time scale.

Hubble's constant is deduced from measurements of redshifts and distances to remote galaxies. There was never any doubt about the redshifts, but the distance scale of the Universe rested on very shaky ground, right back to those few measurements by direct trigono-metrical methods which gave a handful of distances to a few Cepheids which were thereby calibrated as cosmic distance markers. The whole distance scale for objects beyond our Milky Way was dramatically revised upward (and with it the estimate of the age of the Universe) as a result of Hale's legacy of large telescopes, the presence of a German-born astronomer in Los Angeles in the 1940s, and America's involvement in World War Two.

Walter Baade was born in Shröttinghausen, in Germany, in 1893. He was the son of a schoolteacher, and worked his way up through the academic system to emerge from Göttingen University with a PhD in 1919. Right through the 1920s, the decade when both observation and theory began to reveal the true nature of the Universe in which we live, Baade was working in the Bergedorf Observatory, part of Hamburg University. But in 1931 the changing political climate in Germany led Baade, like so many of his contemporaries, to depart for the United States, where he worked at the Mount Wilson and Palomar Observatories for twenty-seven years, returning to Germany in 1958. He died at Göttingen in 1960. In the middle of his sojourn in California, Baade was touched by one of life's little ironies. As a German national, when the US entered the war he was not considered a suitable person to participate in the war effort directly, and while most of his astronomical colleagues were inducted into military research, he was left in more or less splendid isolation with unlimited access to the then-biggest telescope in the world, the 100-inch (2.5-metre). With Los Angeles blacked out under wartime restrictions, and nobody else breathing down his neck for a turn on the instrument, in 1943 Baade was able to push the telescope to its very limits, taking photographs which resolved stars in the inner part of the Andromeda galaxy into individual points of light, where Hubble had only been able to see a fuzzy haze. The observations showed Baade that there were two very different types of star in our neighbouring galaxy. The first type, which he called Population I, are young stars, many of them hot and blue, which are found in the spiral arms. The other type, found on the central part of the galaxy and in the globular clusters of the halo, are older, and, by and large, cooler and redder. He called them Population II. We now know that this is a typical pattern, found in all spiral galaxies, including our own. Population I stars are young; Population II stars are old. And there are other important differences between the populations.

When the lights of Los Angeles came back on, and his colleagues returned from wartime service, Baade was able to continue this work, because by 1948 the 200-inch (5-metre) telescope was in operation. Its greater size more than compensated for the deterioration in viewing conditions. Soon, Baade showed that each population of stars in the Andromeda galaxy had its own type of Cepheid variable. Both Population I Cepheids and Population II Cepheids had well-defined period-luminosity relations – but their two period-luminosity relations were different from one another. The period-luminosity relation used by Hubble was the right one for Population II Cepheids, like those in the halo of our Galaxy. But it had been applied to the hotter and brighter blue Population I Cepheids of the Andromeda galaxy, in ignorance of the fact that these stars had a different period-luminosity relation. The Population I Cepheids were much brighter than their Population II counterparts (which is why Hubble was able to resolve them at all), and when Baade recalculated the distance to the Andromeda galaxy using the correct period-luminosity relation he came up with a figure of two million light years instead of the 800,000 light years Hubble had estimated. The Andromeda galaxy was both brighter and more distant than Hubble had realized.

Because the distance to the Andromeda galaxy had been a crucial step in Hubble's estimate of the scale of the Universe, at a stroke this more than doubled the distance estimated to every external galaxy, and reduced Hubble's constant to less than half its previously calculated value. If they were much further away than had been thought, the galaxies beyond the Milky Way must be that much bigger than had been calculated, in order to appear as large as they did through telescopes on Earth. Indeed, they had to be about the same size as our own Galaxy – some of them bigger still. Instead of our Galaxy seeming to be a giant in the Universe, it now appeared as very average, much the same size as the other galaxies. Newspaper headline writers had fun with the announcement of Baade's results, making much of the fact that the 'size of the Universe' had 'doubled'. But it was of far more importance to cosmologists that the calculated *age* of the Universe had also more than doubled, from two billion to five billion years. At least the Universe now seemed, as of the early 1950s, to be older than the Earth and Solar System.

Over the following thirty years, estimates of the distance scale and age of the Universe were almost continually being refined, as more observations were made. Baade's original estimate itself proved to be distinctly on the low side. Continuing work with the 200-inch, notably by Allan Sandage, an American-born astronomer who joined the team at the Hale Observatories in 1952, suggests that the Universe could be as much as 20 billion years old. There is still a range of uncertainty in all these estimates – remember the tenuous chain of distance estimates which links our measurements in the backwoods of one ordinary galaxy with the far reaches of the cosmos.[11] But few, if any, astronomers today would argue with an estimate that sets the age of the Universe – the time since the Big Bang – as somewhere between 13 and 20 billion years. There is still room for disagreements within that range. Some evidence suggests that the oldest known stars are themselves about 20 billion years old, but some of the studies of the dynamics of the Universe are interpreted as indicating an age of 15 billion years or less. This conflict is important, as we shall see, in trying to decide the ultimate fate of the expanding Universe. But it is largely irrelevant to the search for the Big Bang. In the 1930s, the supposed age of the Universe flatly contradicted the simple Big Bang models; today, such disagreement as there is amounts to discussions over the fine tuning. By and large, the observations and calculations of the Hubble constant agree very well with the simplest kinds of model universes that can be constructed using the Friedman–Lemaître equations.

A choice of universes

Our Universe is expanding and was more dense in the past than it is today. This has now been established, not only by interpreting the redshift of distant galaxies as an expansion effect, but by counting the number of radio galaxies at very high redshift and comparing them with the numbers found in equivalent volumes of space at low redshift. Because light takes time to travel across (or through) space to us, we see distant portions of the Universe (those at high redshift) as they were long ago. If the redshift is so great that light has been 5 billion years on the journey, in effect we see the galaxies from which the light has come as they were 5 billion years ago. So there is direct evidence that the Universe is *both* expanding *and* came from a higher-density state – the simple Steady State model of cosmology is ruled out. This also means that we can rule out of consideration all the simple collapsing models that are allowed for by the equations. The only models that could possibly be relevant to the real Universe are the ones which include at least a phase of expansion. That still leaves room for some rather strange models, including some in which the model universe begins life in a state of very low density, contracts for a long time, slows down and then expands once again. But because, as we shall see in the next chapter, there is now very good evidence that the Universe as we know it started out from a very dense, very hot state I shall ignore these exotic ideas as well. (Of course, if this particular variation on the theme contracts so much that it reaches a state of very high temperature and density before expanding once again, then it becomes, from the point of view of all human observations, a Big Bang model).

The very dense, very hot state in which the Universe was born is commonly called the Big Bang. Fred Hoyle seems to have been the person who introduced the term to astronomy, in a scientific paper he published in 1950. Many astronomers don't like the term, because it is in some ways misleading. It gives the impression of an explosion occurring in the middle of empty space, like a giant firecracker, or a scaled-up nuclear bomb. But when space itself is part of the expansion, and matter is just carried along by the expanding space, the bomb analogy breaks down. Even when the Universe was very dense, it was very smooth. There was no difference in pressure forcing it to expand, and there were no sound waves to make the 'bang' audible. The expansion was a smooth affair, which continues to this day. But, though the purists may, rightly, bemoan the fact, we are stuck with the name Big Bang now, and I shan't try to swim against the tide.

The variety of possible universes depends on the curvature of spacetime. Positive curvature can give the equivalent of a closed surface, like that of a sphere, but even a surface with positive curvature may not be closed in this way; it could extend off to infinity. Perhaps this is easier to understand in terms of lines, one-dimensional objects. A line may be closed back on itself, like a circle, so that it has a definite size, the circumference of the circle. Or it may be open, like a hyperbola, which extends out to infinity on either side of the bend which gives it its distinctive shape. The Universe we live in may be closed, the equivalent of a circle (or a sphere) and with a finite extent. Or it may be open, like a hyperbola or a vast hyperbolic bowl, stretching out to infinity and containing an infinite amount of matter. What decides the amount of curvature, and thereby whether or not the Universe is open or closed, is the amount of matter in the Universe, because gravity, produced by matter, is what curves spacetime. This is important in deciding the ultimate fate of the Universe, as we shall see in Chapter 10, but both the open and closed variations on the theme start out in very much the same way.

Negative curvature is equivalent to a type of curved surface which can *only* be open, like a saddle surface. The amount of the curvature itself depends once again on how much matter there is in the Universe (or, more accurately, on the density of matter at any time in its evolution, any 'cosmic epoch'). Einstein's cosmological constant provides more room for variety, but I will mention only one of those extra possibilities.

The best way to get a handle on all of this is to look at a property the cosmologists call the 'scale factor', and see how it changes as a particular model universe evolves. The scale factor can be thought of as the separation between a chosen pair of galaxies – it increases as the universe expands, and is denoted by the letter R. When R doubles, the distance between every pair of galaxies doubles, and so on. Pushing the observed behaviour of the real Universe – its expansion – back to the point where everything was touching everything else, R starts out from zero, equivalent to the birth of the universe in an infinitely dense state. If we plot a graph with the scale factor R against the time since the Big Bang, t, we can see how the main types of model universe differ from one another. What matters is the extent to which *space* is curved, and how the curvature of space (not spacetime) changes as time passes. In all cases (assuming that the cosmological constant is zero) the expansion of the universe slows down as t increases. This can easily be understood in terms of the effect of gravity, holding back the expansion of the universe. But the rate at which the slowdown occurs depends on how much matter there is in the universe, which decides whether or not it is open or closed.

The expansion of the universe slows down as time passes, so that R increases by a smaller amount in later time intervals that are the same size. This slowdown occurs most rapidly in the closed universes.

For all the universes with positive curvature, the R curve bends determinedly to the right, eventually turning over and becoming a description of a collapsing, not an expanding, universe. Such a universe contains so much matter that gravity eventually causes it to recollapse. With ever increasing speed, such a universe then rushes back into a state of infinite density, like the Big Bang in which it was born. The space in such a closed universe is closed, in the same sense that the surface of a sphere is closed; the change in curvature as time passes is equivalent to the sphere being first inflated and then deflated. In a universe with negative curvature, the expansion continues forever, even though it is always slowing down, and the universe, like a saddle surface, is infinite. The space of such a universe is called hyperbolic, in the same way that the space of a closed universe is termed spherical.[12] There are whole families of closed universe models, and whole families of open universe models; the Einstein–de Sitter universe is a unique special case exactly balanced between models with positive curvature and those with negative curvature.

Having been stuck with the name Big Bang for the birth of the Universe, cosmologists have at least been consistent with their terminology for its possible end. A universe that recollapses ends in a 'big crunch', sometimes referred to as the 'Omega Point', while the ones that expand forever end, logically enough, in what T. S. Eliot told us was the opposite of a bang: 'This is the way the world ends, not with a bang but a whimper.' So there are two basic alternatives, the bang-crunch universes and the bang-whimper universes. Everything else is just fine tuning.[13]

Figure 5.4 gives an indication of how these different types of universe evolve. The figure also shows how our estimates of the age of the real Universe in terms of Hubble's parameter, H, must be only an approximation. By setting the age of the Universe as $1/H$, we assume that the expansion has always continued at the same rate – in effect, we draw a tangent to the R/t curve and extrapolate it back to find when it crosses the axis, the time when R was zero. If the expansion is slowing down, in a space of positive curvature, then this estimate, $1/H$, is always *more* than the true age of the Universe. This made the contradiction between the Hubble age for the Universe and the estimated age of the Earth and stars even more embarrassing in

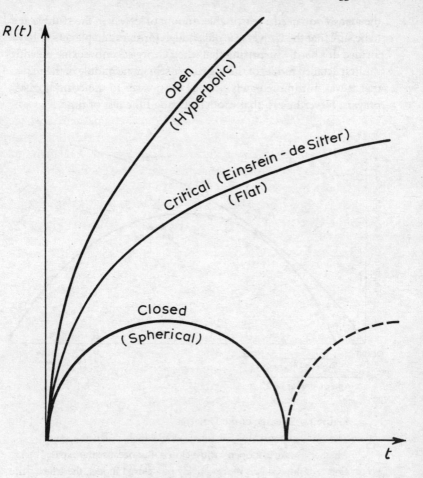

Figure 5.4 The three geometries of space described in Figure 5.2 correspond to three types of Universe which all start from a compact state and expand, at least initially, as time passes. R is the scale factor of the universe; t represents time. Both the open and the flat universes expand forever; the closed universes collapse eventually, and may then undergo further cycles of expansion and collapse.

the 1940s – you need a reasonable amount of leeway in the Hubble age to be sure that the Universe is old enough for stars and planets to have formed. It's hardly surprising that when George Gamow came up with the first detailed model of the Big Bang itself, in the middle of the 1940s, that it was not immediately seen as the answer to the cosmologists' prayers. Nevertheless, that model has stood the test of time.

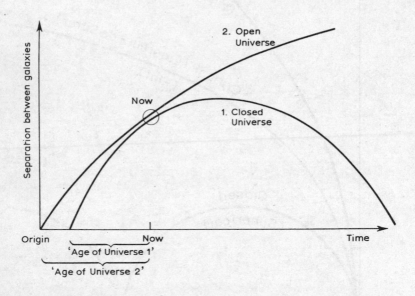

Figure 5.5 The Age of the Universe
The state of the Universe today would look much the same whether it were just open or just closed. But because the expansion rate slows down more quickly in a closed model, the true age of the Universe is much less, if it is closed, than indicated by simple measurements of the way galaxies are moving apart. The simple measurements give an 'age' of about 20 billion years; the true age is probably about two-thirds of that figure.

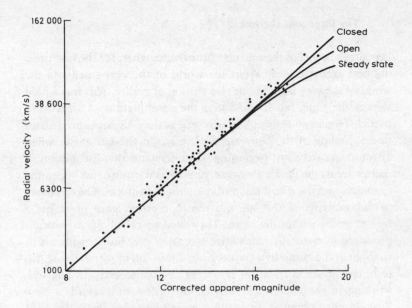

Figure 5.6 Observations of distant galaxies could, in principle, tell us what kind of universe we live in. By looking far out to space, we see galaxies as they were when the Universe was younger and expanding more rapidly – when 'Hubble's constant' was different. So the straight line in Hubble's diagram should be bent slightly, by an amount which indicates how quickly the expansion of the Universe is slowing down. Unfortunately, the range of our telescopes scarcely begins to provide enough evidence to decide which of the models of Figure 5.4 is the best bet. Although it does look as if the Steady State model is ruled out, there is no way in which the available observations (collected over the years by Allan Sandage) can distinguish whether our Universe is open or closed. From this evidence, our Universe is seen to be very close to the special, in-between case – a flat spacetime. In this diagram, magnitude effectively corresponds to distance, and radial velocity to redshift; it is the modern, vastly extended, version of Hubble's diagram in Figure 4.1.

The large and the small

Lemaître's model of the universe brought together, for the first time, the best available ideas about the world of the very small and the world of the very large in one description of reality. This was a bold step at the time, but in the 1980s the combination of ideas from particle physics and cosmology was seen as the only route towards an understanding of the Universe we live in. In the late 1920s, when astronomers were just beginning to understand the distribution of matter across the visible Universe, physicists were also just beginning to understand the distribution of matter *within* atoms. The discovery of radioactivity, in the late nineteenth century, gave physicists a tool to probe within the atom. They used so-called alpha particles, produced by naturally radioactive atoms, as tiny bullets with which to shoot at the atoms in a crystal, or in a thin foil of metal. Using this technique, researchers at the University of Manchester, in England, working in the department headed by Ernest Rutherford, a New Zealand-born physicist, found that most often alpha particles went right through a thin metal foil target, but that occasionally a particle would be bounced back almost the way it came. Rutherford came up with an explanation for this behaviour in 1911, and gave us the basic model of the atom that we learn about in school today.[14]

Rutherford realized that most of the material of an atom must be concentrated in a tiny inner core, which he called the nucleus, surrounded by a cloud of electrons. Alpha particles, which come from radioactive atoms, are actually fragments of atomic nucleus. When such a particle hits the electron cloud of an atom, it brushes its way through almost unaffected. But electrons carry negative charge, while atoms as a whole are electrically neutral. So the positive charge of an atom must be concentrated, like its mass, in the nucleus. Alpha particles too are positively charged. And when an alpha particle hits an atomic nucleus head on, the repulsion between like charges halts it in its tracks and then pushes it back from where it came. Later experiments confirmed the broad accuracy of Rutherford's picture of the atom. Most of the mass and all the positive charge is concentrated in a nucleus about one hundred thousandth of the size of the atom.

The rest of the space is occupied by a tenuous cloud of very light electrons that carry negative charge. In round numbers, a nucleus is about 10^{-13} cm across, while an atom is about 10^{-8} cm across. Very roughly, the proportion is like a grain of sand at the centre of Carnegie Hall. The empty hall is the 'atom'; the grain of sand is the 'nucleus'.

The particle that carries the positive charge in the nucleus is called the proton. It has a charge exactly the same size as the charge on the electron, but with opposite sign. Each proton is about two thousand times as massive as each electron. In the simplest version of Rutherford's model of the atom, there was nothing but electrons and protons, in equal numbers but with the protons confined to the nucleus, in spite of them all having the same charge, which ought to make them repel one another (like charges behave in the same way as like magnetic poles do in this respect). There must be another force, which only operates at very short ranges, that overcomes the electric force and glues the nucleus together – more of this in Chapter 9. But over the twenty years following Rutherford's proposal of this model of the atom, a suspicion grew up among physicists that there ought to be another particle, a counterpart of the proton with much the same mass, but electrically neutral. Among other things, the presence of such particles in the nucleus would provide something for the positively charged protons to hold on to without being electrically repulsed. And the presence of neutrons, as they were soon called, could explain why some atoms could have identical chemical properties to one another but slightly different mass. Chemical properties depend on the electron cloud of an atom, the visible 'face' that it shows to other atoms. Atoms with identical chemistry must have identical numbers of electrons, and therefore identical numbers of protons. But they could have different numbers of neutrons, and therefore different masses. Such close atomic cousins are now called isotopes.[15]

The great variety of elements in the world are all built on this simple scheme. Hydrogen, with a nucleus consisting of one proton, and with one electron outside it, is the simplest; the most common form of carbon, an atom that is the very basis of living things, including ourselves, has six protons and six neutrons in the nucleus of each atom, with six electrons in a cloud surrounding the nucleus. But there are nuclei which contain many more particles (more nucleons) than this. Iron has 26 protons in its nucleus and, in the most common isotope, 30 neutrons, making 56 nucleons in all, while uranium is one of the most massive naturally occurring elements, with 92 protons and no less than 143 neutrons in each nucleus of uranium-235, the radioactive isotope which is used as a source of nuclear energy. Energy can be obtained from the fission of very heavy nuclei because the most stable state an atomic nucleus could possibly be in, with the least energy, is iron-56. In terms of energy, iron-56 lies at the bottom of a valley, with lighter nuclei, including those of oxygen, carbon, helium and hydrogen, up one side and heavier nuclei, including cobalt, nickel, uranium and plutonium, up the other side. Just as a ball which lies on the valley's sloping side can more easily be kicked down into the bottom of the valley than higher up the slope, so if heavy nuclei can be persuaded to split, they can, under the right circumstances, form more stable nuclei 'lower down the slope', with energy being released; equally, if light nuclei can be persuaded to fuse together, then they too form a more stable configuration with energy being released.[16] The fission process, which Lemaître tried to extend to the primeval atom, is what powers an atomic bomb; the fusion process, which Gamow applied in his model of the Big Bang, is what provides the energy from a hydrogen, or fusion, bomb, in which hydrogen nuclei are converted into helium nuclei. But all that still lay in the future in the 1920s. Although there was circumstantial evidence for the existence of neutrons in that decade, it was only in 1932 that James Chadwick, a former student of Rutherford who was by then working at the Cavendish Laboratory in Cambridge (where Rutherford was the Director), carried out experiments which proved that neutrons really existed.

So when Lemaître first proposed his 'primeval atom' model of the origin of the universe, nobody actually knew what a real atom was like. The term primeval atom is itself a misnomer, and primeval nucleus would be much better. When we talk of atoms 'splitting' or undergoing radioactive decay, what we really mean is that their *nuclei* break up into two or more parts, or that one nucleus ejects a particle, such as an alpha particle, and changes into the nucleus of a lighter element. Lemaître envisaged this as the process which, by repeated fission, produced all the matter in the Universe from one 'nucleus'.

But this could not explain (quite apart from more technical problems, which I won't go into) why well over half of the matter in stars and galaxies is in the form of hydrogen, the lightest and simplest element, while most of the rest is helium, the next lightest element. Spectroscopic studies of stars and galaxies unambiguously show the Universe to be dominated by the two simplest elements, hydrogen with just one proton in its nucleus and one electron to go with it, and helium with two protons and two neutrons in its nucleus, and two electrons outside. Splitting all of the primeval nucleus into such simple components is too big a task; but Gamow hit on the idea that the Big Bang might have started out with the very simplest particles and that the Universe then built up the heavier elements by adding more protons and neutrons to the simplest nuclei. After all, if you *start* with hydrogen, then at a stroke you explain the existence of more than half of the nuclei in the Universe.

Gamow's Universe

George Gamow was a larger-than-life character with a boundless imagination which took him from nuclear physics to cosmology and then into the world of molecular biology. He made significant contributions in all three areas of science – *the* three key areas of twentieth-century science – and found time along the way to write books for the layman, carry out elaborate practical jokes on his colleagues, and generally to illuminate the world of science in the middle decades of the century. He achieved all this despite being indifferent about such minor details of life as spelling or dates, and being hopeless at working out simple arithmetic. Born in the Ukraine, at Odessa, in 1904, after he moved permanently to America in the mid-1930s Gamow always signed his letters to his friends 'Geo.', an abbreviation which he was unshakably convinced was pronounced 'Joe'; so 'Joe' he was, to the very large number of those friends, until his death in 1968.

Having lived through the turmoil of revolution and civil war in Russia, in 1922 Gamow enrolled at Novorossysky University, but soon transferred to the University of Leningrad, where he stayed until 1928, gaining a PhD and learning about Friedman's models of the universe from Friedman himself. Once qualified, he travelled to the University of Göttingen, then to the Institute of Theoretical Physics in Copenhagen, then to the Cavendish Laboratory in Cambridge, and then back to Copenhagen. The three scientific centres he visited in the years from 1928 to 1931 were at the heart of the revolution in physics then taking place, the discovery of quantum physics and the beginnings of the application of the new theory to an understanding of atoms. Gamow learned his quantum physics from the pioneers in the subject, just as he learned his cosmology from one of the pioneers. And during his visit to Göttingen he made the first of his major contributions to science, applying quantum theory to explain how an alpha particle could escape from an atomic nucleus.

Each of these alpha particles, it is now known, consists of two protons and two neutrons, held together by the strong nuclear force which overcomes the electric force of repulsion between the protons. They are, indeed, identical to helium nuclei, in effect helium atoms from which the two electrons (all that the atom possesses) have been removed. When an alpha particle is inside the nucleus of a very heavy atom, it is held in by the strong nuclear force. If an alpha particle is just outside the nucleus, however, the electric repulsion force dominates, because the nuclear force has only a very short range, and so the particle is ejected. Extending the idea of a stable state as being at the bottom of a valley, in energy terms, for the alpha particle the nucleus is like the interior of an extinct volcano. Deep in the heart of the volcano, it is energetically stable; but if it were just a little bit outside the volcano it would be on the steeply sloping sides of the mountain, and would rapidly roll away. Gamow showed how an alpha particle could get over the hill, as it were, from just inside the nucleus to just outside – and his explanation of alpha decay was the first successful application of quantum theory to the nucleus.

In 1931, Gamow was called back to the USSR, where he was appointed Master of Research at the Academy of Sciences in Leningrad, and Professor of Physics at Leningrad University. But his ebullient nature and independence of mind hardly suited him to a happy life under Stalin's regime in the 1930s, and when he was allowed to attend a scientific conference in Brussels in 1933 he seized the opportunity

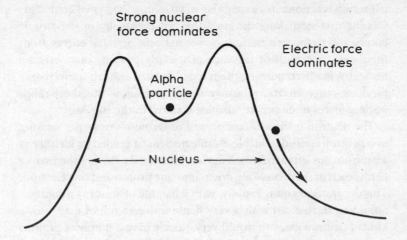

Strong nuclear
force dominates

Electric force
dominates

Alpha
particle

Nucleus

Figure 5.7 An alpha particle inside a nucleus is trapped by the strong nuclear force. The same particle just outside the nucleus would 'roll away' as it is pushed by the electric force. But how can an alpha particle get out of the nucleus without enough energy to climb the intervening hill? George Gamow's explanation of this alpha decay process was the first successful application of quantum physics to the nucleus.

to stay away, moving to George Washington University in Washington State, where he was Professor of Physics from 1934 to 1956, and then to the University of Colorado, in Boulder, where he stayed until his death.

Gamow's interest in how things got *out* of atomic nuclei led him to wonder about the possibility of particles getting *in*, climbing the hill from outside and dropping into the region where the strong nuclear force dominates. He was involved in pioneering calculations which showed that if protons could be fired into atoms with energies of a few hundred kilovolts then they would trigger the kind of nuclear reactions that cause nuclear fission and alpha decay. John Cockroft and Ernest Walton, working at the Cavendish Lab, did just that in 1932, creating the world's first particle accelerator, using a high-voltage electric field to accelerate protons and smash them into atoms, and triggering reactions in exactly the way Gamow had predicted. This was the first step along the road that led eventually to the atomic bomb, and to the first nuclear power stations, deriving energy from the fission process. But the idea of sticking protons onto existing nuclei by, in effect, pushing them together hard enough to overcome the long-range electric repulsive forces and allow the short-range nuclear forces to dominate, also led Gamow to the Big Bang.

The neutron is the key component of Gamow's universe. As long as a neutron is inside a stable atomic nucleus, it retains its identity as a neutron. But left to their own devices individual neutrons themselves decay, each of them breaking down into one proton and one electron. This decay occurs quite rapidly, with a half life of about 13 minutes.[17] So, if you started out with a very dense universe full of neutrons, a kind of neutron gas, you would very quickly have a supply of protons and electrons as well, with precisely enough particles of each kind around (the same number of electrons as protons) for every stable atom that might be formed to have an equal number of protons and electrons and leave no excess electric charge left over.

Gamow's idea immediately supplied hydrogen for the universe. Each hydrogen atom consists simply of one proton with one electron held in its neighbourhood by the electric force of attraction between opposite charges. Allow a neutron to decay, and there you are with a bare hydrogen nucleus and a handy electron, all ready to make an atom. But where do the rest of the atoms, those corresponding to helium and the heavier elements, come from?

During the 1940s, Gamow was joined at George Washington University by Ralph Alpher, a graduate student to whom he assigned the task of working out the details of how more complex nuclei might have been built up from hydrogen (a process known as nucleosynthesis) in the Big Bang. The model they developed depended on collisions between the particles in the cosmic soup of dense material in the first few minutes of the life of the universe.[18] The calculations showed that it would be relatively easy for a proton (hydrogen nucleus) and a neutron to collide strongly enough to overcome electrostatic repulsion and stick together, to form a nucleus of deuterium, also known as heavy hydrogen. Another collision with a neutron would produce a nucleus of tritium, containing one proton and two neutrons. But tritium is unstable, so that one of its neutrons soon spits out an electron and becomes a proton. The nucleus has now evolved into one that corresponds to an isotope of helium, containing two protons and one neutron, and called, for obvious reasons, helium-3. All it needs now is for another neutron to stick to the growing nucleus to make an alpha particle, the nucleus of a helium-4 atom. So far, so good. There was no need to worry about the electrons, since once the nuclei were manufactured they could easily pick up the electrons they needed from the swarm of particles in the primeval soup. But at that point the model ran into a snag.

The helium-4 nucleus, the alpha particle, is a particularly stable state. It is very disinclined either to break up into smaller components, or to accept additional components and grow into something more complex. Worse, there is no naturally occurring element which has a nucleus containing five particles, and when such a nucleus is made artificially in the lab by bombarding helium-4 with neutrons it immediately breaks down to helium-4 again. In order to get around this difficulty, Gamow and Alpher had to speculate that a single helium-4 nucleus might occasionally be struck *simultaneously* by two particles and capture them both to form a nucleus containing six particles. Even if this happens, the same problem arises for the nucleus containing eight particles, which very rapidly breaks down into two alpha particles.[19] And with the universe rapidly thinning out as it expands away from the superdense state of the Big Bang, by the time you have made the helium the chance of a double collision of this kind is small, and rapidly getting smaller. In the 1940s, although the prospect of getting over these gaps by the capture of two particles at once seemed unlikely, there was just enough ignorance about conditions in the early Universe, and about the rates at which such nuclear reactions might occur, to allow Gamow and Alpher to get away with the idea, as a working hypothesis. After all, as Gamow used to tell anyone who was interested, the theory explained where all of the hydrogen and all of the helium in the Universe came from, and that accounted for more than 99 per cent of the matter visible in stars and galaxies. Even if the theory didn't properly explain the synthesis of the heavy elements (to an astronomer, anything except hydrogen and helium is a 'heavy' element) they represented less than one per cent of the problem.

The detailed calculations of how nuclei can capture neutrons or protons (the numbers that come out of the calculations are called capture cross sections) formed the basis of Alpher's PhD thesis, submitted in 1948. It clearly deserved a wider audience, however, and Alpher and Gamow wrote up a paper for submission to the *Physical Review*. At this point, Gamow's sense of fun overcame him, and he perpetrated his most famous scientific joke. 'It seemed unfair to the Greek alphabet,' he wrote later,[20] 'to have the article signed by Alpher and Gamow only, and so the name of Dr Hans A. Bethe (*in absentia*) was inserted in preparing the manuscript for print. Dr Bethe, who received a copy of the manuscript, did not object.' So the classic paper in which the modern version of the Big Bang model first saw the light of day appeared, on 1 April 1948 (a coincidence which delighted Gamow) under the names Alpher, Bethe and Gamow. To this day, it is known as the 'alpha, beta, gamma' paper, a suitable reflection of the fact that it deals with the beginning of things, and also of the importance of particle physics to cosmology (alpha particles we have already met; beta particle is another name for electron, and gamma ray is the name for an intense pulse of electromagnetic radiation, an energetic photon).

This early version of the Big Bang appeared in the same year, 1948, that Fred Hoyle, Tommy Gold and Hermann Bondi came up with their idea of an expanding Steady State Universe. Right through the 1950s and into the 1960s the two rival ideas stirred debate among the experts, with Hoyle as the leading Steady Stater and Gamow the leading Big Banger in friendly rivalry. Ironically, it was to be Hoyle who would show how to resolve the greatest difficulty with Gamow's universe, finding a way to make the heavy elements inside stars, once the initial job of cooking up helium in the Big Bang had been carried out. But there is an even greater irony in the story, involving one of the most significant missed opportunities in the history of science, and emphasizing the way even cosmologists failed to take their equations seriously at that time.

Two questions

In those days, cosmology was very much a game. Rival models were developed and tested against one another almost as a kind of abstract mathematical duel, with little thought that one of these models might, in fact, be a correct mathematical description of our own Universe. Even Gamow, who loved his theory of the Universe as if it were his own son, fell into this trap.

The conditions needed to make even helium in the Big Bang (never mind the heavy elements for now) include both very high density *and* very high temperature. Although you might imagine a cold soup of neutrons expanding away from a state of very high density, quite simple calculations show that such a cold neutron soup is very rapidly converted almost entirely into helium. It is only in a hot Big Bang that most of the matter stays as hydrogen, and, curiously but just one of those things, it doesn't make much difference what the exact density of the model universe is a few seconds after the moment of creation. Provided the universe is hot, you always end up with about a third of the matter being turned into helium, with the rest staying as hydrogen until it is reprocessed in stars as the Universe evolves.

Most of the hydrogen is prevented from being cooked into helium while the Universe is very dense by the presence of a great deal of energetic radiation. This electromagnetic radiation can be thought of in terms of particles, called photons. Alpher and another young researcher, Robert Herman, used the fact that about a third of the Universe is helium and the rest hydrogen to calculate how many photons there must be in the Universe; it comes out at a staggering billion photons for every nuclear particle (that is, for every proton or neutron). Radiation – photons – is a form of energy and the density of the radiation (the amount of energy in a certain volume of space) can be expressed in terms of temperature. Winding back the Friedman equations to the early seconds of the Universe, Alpher and Herman showed that there must have been a time when the energy density of the radiation was greater than the energy density of the matter, given in terms of Einstein's famous equation $E = mc^2$. The Gamow universe was born out of a fireball of radiation, quickly cooling as it expanded, and becoming dominated by matter only after it had expanded, and cooled, by a critical amount. But the radiation would still be there, filling the entire Universe but getting thinner, cooler and weaker as time went by. In 1948, Alpher and Herman published a paper in which they calculated that the temperature of this leftover radiation today must be about five degrees Absolute, 5 K.[21]

In his popular book *The Creation of the Universe*, published in 1952, Gamow gives a slightly different estimate for the temperature of the Universe today (and also bemoans Herman's stubborn refusal to change his name to Delter). He derives an equation which says that the temperature is equal to 1.5×10^{10}, divided by the square root of the age of the Universe in seconds. This gave him an estimate of about 50 K. At other times in the early 1950s, Gamow and his colleagues came up with other figures in the range from 5 K to 50 K, depending on what assumptions they made about the state of the early Universe and its age. Today, particle physicists calculate that a more accurate version of Gamow's equation simply sets the temperature now as 10^{10} divided by the square root of its age in seconds, and in addition the estimates for the age have increased, all of which reduces the top estimate for the temperature of the Universe at present. This equation is only an approximate relation, and there are better ways to calculate the temperature of the Universe at any epoch. But it is a useful rule of thumb, which tells us, for instance, that one second after the moment of creation the temperature was 10 billion degrees, that after 100 seconds it had already cooled to one billion degrees, and that after an hour it was down to 170 million degrees. For comparison, the temperature at the heart of our Sun is calculated to be about 15 million degrees.

Here was a clear prediction made by the hot Big Bang theory. It said that the Universe ought to be filled with a sea of radiation with an energy equivalent to a temperature of a few K. Such radiation would be detectable at radio wavelengths, and radio astronomy was just getting started in the early 1950s. But no radio astronomer picked up the idea and went out to test it, while Gamow and his team went their own ways into other areas of research (Gamow himself becoming fascinated by the problem of cracking the genetic code of DNA), and never went out on the campaign trail to drum up interest in the idea and encourage, or force, the radio astronomers into appropriate action. What went wrong? The best explanation is the one put forward by physicist Steven Weinberg in his book *The First Three Minutes*. 'It was,' he said, 'extraordinarily difficult for physicists to take seriously *any* theory of the early universe' in those days. 'Our mistake is not that we take our theories too seriously, but that we do not take them seriously enough. It is always hard to realize that these numbers and equations that we play with at our desks have something to do with the real world.'[22]

By 1956, when Gamow went on his way to Colorado and the team disbanded, the early version of the hot Big Bang model had posed two questions, the answers to which were to provide the basis for further developments. One of those questions was widely recognized, and great strides towards answering it were made in the late 1950s and early 1960s. It was the question of where the heavy elements come from, if they are not manufactured in the Big Bang. The other question was unnoticed and remained buried in the scientific literature until it was answered by accident in 1964. It was the question of the background temperature of the Universe today. The combination of these two answers – each of which led to the award of a Nobel Prize – with Gamow's Universe initiated the modern era in cosmology.

Chapter 6

Two Keys to the Universe

In the 1930s, astrophysicists were much more interested in the workings of the Sun and stars than in the origin of the Universe. The mystery of creation still seemed more a topic for the metaphysicians to muse over than something that fell within the scope of scientific investigation; and the puzzle of how the stars shone was both a fascinating one and something that was just being understood in terms of the revolutionary new understanding of physics that had broken through in the 1920s. But as it turned out, the investigation of how stars stay hot led directly to a better understanding of the Universe at large, and of the Big Bang itself.

Our Sun is the nearest star, and the one we know most about. If astronomers were to have any hope of understanding the workings of stars in general, they had to understand at least the outlines of the workings of our own Sun. But at first, even the new physics, quantum physics, seemed inadequate for the task.

All the evidence that the Earth had been around for four or five billion years clearly implied a similar age for the Sun, and even in the nineteenth century physicists appreciated that no ordinary chemical process of combustion could have kept the Sun hot for that long. A solid ball of coal, for example, the size of the Sun and burning in an atmosphere of pure oxygen to generate the same amount of heat every second that the Sun is putting out today, would have burned to a cinder in about 1,500 years. The first attempts to explain how stars could stay hot for a very long time by invoking astrophysical processes, processes that operate in the stars but not here on Earth, came in the second half of the nineteenth century, from the German physicist Herman Helmholtz and the British physicist William Thomson (later Lord Kelvin, in whose honour the absolute scale of temperature is named).

Helmholtz and Thomson were both major figures in the scientific world of the time, with wide-ranging interests. Their paths crossed through a common interest in the age of the Earth, and therefore the age of the Sun. In 1854, Helmholtz came up with an age of 25 million years; a little later, Thomson came up with a slightly longer estimate, giving 100 million years as the most likely figure. Even Thomson, we now know, was more than ten times too modest in his assessment of the age of the Solar System. But these estimates were a significant step away from the idea of a creation only a few thousand years ago, an idea still accepted by some Church authorities at that time (Darwin's *Origin of Species* was published in 1859; the estimates of the age of the Earth made by Helmholtz and Thomson were very far from being esoteric scientific data, and were directly relevant to the greatest scientific and philosophical debate of the time). But where could the energy come from to keep the Sun hot for even a hundred million years?

The answer seemed to be 'gravity'. If the Sun had started out as a thin cloud of gas in space, falling together into a more compact ball under the tug of its own gravity, then it would have warmed up as it collapsed. If you stretch a spring, you have to do work (put in energy) to overcome the elastic forces in the spring; when you let go of the spring, the energy is released. The same sort of thing happens if you lift a weight up from the ground. You are putting energy – in this case, in the form of gravitational potential energy – into the weight. When you drop the weight, that energy is turned into kinetic energy as the weight falls, and then the kinetic energy is turned into heat energy when the ground stops the weight's fall. All the particles that make up the Sun would 'like' to fall to its centre, to the centre of mass of the system which they are part of. If they could do so, they would give up gravitational potential energy, ultimately in the form of heat, in the same way that a falling weight gives up energy as heat. A more compact star is at a lower energy state than a more diffuse star, because its component particles are closer to the centre of mass.[1] So if you start out with a star like the Sun, but slightly bigger, and leave it to contract slowly under the tug of its own gravity then you would expect heat to be generated.

The age of the Sun

Astronomers are well able to calculate the distance to the Sun (the astronomical unit of distance) and its mass from the orbital motions of bodies in the Solar System, the strength of tides on Earth, and so on. They know how much energy it radiates each second, because they know how bright it must actually be to look as bright as it does in our sky. The energy released is, in fact, about 4×10^{33} ergs per second, some 10^{41} ergs per year. When Helmholtz and Thomson calculated how much energy a star like the Sun would release if it contracted slightly, they found that the collapse from a diffuse gas cloud into the star would provide enough energy to keep the star radiating this strongly for 10 or 100 million years, before the interior of the star must cool sufficiently to trigger another major collapse. Somewhat unfairly on Helmholtz, both alphabetically and in chronological order of making the calculations, this time scale is known today as the Kelvin–Helmholtz time scale. It is undoubtedly important in the early stages of the life of a star, and the gravitational heating effect is the process which makes stars hot in the first place, and causes them to begin to shine. But by the 1920s it was clear that the Earth and the Sun (and therefore presumably the stars) must be much older than Kelvin and Helmholtz had thought – billions of years old, not tens of millions of years old. So where had the energy come from to maintain that 4×10^{33} ergs per second for all that time, and to keep the Sun shining today?[2]

Kelvin's estimate for the age of the Earth was based on measurements of the temperature at the surface of the Earth compared with the temperature down deep mine shafts. Heat is leaking out from the interior, and by estimating how rapidly heat is escaping and working backwards in time he concluded that the whole of our planet must have been molten a few tens of millions of years ago. Kelvin, and the astronomers, found the agreement between this age for the Earth and the Kelvin–Helmholtz age for the Sun persuasive. But even in the late nineteenth century, geologists had clear evidence that the Earth must be much more than a hundred million years old, or there would not have been time for the great thicknesses of folded rocks in structures like the Alps to have formed, while evolutionary biologists also preferred a longer time scale of Earth history during which evolution could have been at work to produce the complexity of life on Earth today. Kelvin's estimate for the age of the Earth was wrong, and the reason became clear at the turn of the century, with the discovery of radioactivity. Radioactive elements occur naturally in common rocks all over the world, and the whole basis for their radioactivity is that their nuclei are moving to a lower energy state by splitting into two or more parts, ejecting alpha particles or whatever. The change in energy of the radioactive nucleus when it decays appears as heat, and in round terms about 90 per cent of the heat flow through the rocks of the Earth's crust is due to radioactivity. So Kelvin's age for the Earth has to be multiplied by something between ten and a hundred to give a more accurate, but still rough, guide to how long the Sun and planets have been in existence.

During the 1920s several developments led to the beginning of a better understanding of how the Sun and stars work. First, the spectroscopists were able to show that our Sun is composed of about 70 per cent hydrogen, 28 per cent helium, and just 2 per cent of heavy elements. Then, Arthur Eddington, inventing the science of astrophysics almost single-handedly, discovered, from studies of binary stars, that the brightness of a star is directly related to its mass. This discovery, announced in 1924, was a key step towards an understanding of how stars work, and in 1926 Eddington published a landmark book *The Internal Structure of the Stars*, which showed how this, and other properties of stars, could be explained in terms of the basic laws of physics describing the behaviour of a large, hot ball of gas. At the same time, continued probing of the atom and investigation of radioactivity, following on from Rutherford's pioneering work, had begun to make physicists familiar (or, at least, less unfamiliar) with one of the more startling predictions of Einstein's special theory of relativity, which had been published back in 1905.

This was the prediction that energy and mass are equivalent and interchangeable, that $E = mc^2$. As physicists became able to measure the masses of atoms (and therefore of their nuclei) more accurately, and as their understanding of the forces which hold nuclei together slowly developed, it became clear that when a heavy nucleus splits into two lighter components the energy that is liberated is exactly compensated for by a loss in mass of the components. When nucleus A splits into B and C, the mass of B and C added together is *less* than the mass of A, by just the amount needed to provide the energy released by the fission, from the rule $E = mc^2$. In the 1920s, and then increasingly in the 1930s, physicists, armed with the new quantum theory as well as relativity, began to come to grips with the implications of Einstein's theory for the world of nuclei.

The same rule, balancing energy gained and mass lost, applies to nuclei on the other side of the valley of stability. But in that case, of course, it is the *addition* of two lighter components to make a heavier nucleus that produces a loss of mass and a corresponding release of energy. A nucleus of helium-4, for instance, consists of two protons and two neutrons. Physicists measure the masses of atoms and nuclei in terms of the mass of an atom of carbon-12, which contains six protons and six neutrons (plus six electrons in a cloud outside the nucleus). This mass is defined as 12 atomic mass units. On this scale, the mass of a proton is 1.007275, and the mass of a neutron is 1.008664. Take two protons and two neutrons, and you 'ought' to get a total mass of 4.031878. But the mass of an alpha particle, a helium nucleus, is 4.00140 atomic mass units. A little more than 0.03 atomic mass units have been 'lost', and appear as heat energy every time the four constituent particles combine together to make a helium nucleus. The amount of mass converted into energy is about 0.75 per cent of the total you started with, and since this mass has to be multiplied by the square of the speed of light, and the speed of light is 3×10^{10} cm per second, that becomes very significant as soon as you have a means to manufacture helium nuclei inside stars.[3]

In the late 1920s and early 1930s, Eddington and his colleagues did not have a detailed theory of just how four hydrogen nuclei could be converted into one helium nucleus. But they knew that mass would be lost in the process, converted into energy, and they knew how much energy the Sun was radiating every second. Using Einstein's equation, it is simple to calculate that the present brightness of the Sun can be maintained by converting just over four million tons of matter into energy every second. That is the amount of matter converted into helium (alpha particles). It sounds a lot, but it is a tiny fraction of the Sun's mass. In round numbers, the mass of the Sun is 2×10^{33} grams, or 2×10^{27} tons. Converting 600 million (6×10^8) tons of hydrogen into helium each second, the Sun 'burns' just under 2×10^{16} tons of fuel each year. In a thousand years, it burns 2×10^{19} tons of hydrogen; in a million years, 2×10^{22} tons. And even after 10 billion years of hydrogen burning at this rate it has used only 2×10^{26} tons, just over 10 per cent of the star's total mass. About 7 per cent of this fuel, more than 14×10^{24} tons of matter, is converted totally into energy in the process. So it would take, in round terms, 10 billion years of 'burning' hydrogen into helium at this rate to change the composition of the Sun to the point where its visible appearance would alter significantly. Here, immediately, is the reconciliation between the age of the Earth and the age of the Sun. If the Earth is four or five billion years old, and a star like the Sun can burn hydrogen steadily for about 10 billion years, then we are scarcely halfway through the lifetime of the Sun in the form that we know it.

Although the broad outlines of the theory looked very good indeed, when Eddington proposed that fusion of hydrogen into helium must be the power source of the Sun and stars many of his physicist contemporaries dismissed the idea. Their reason for doing so made sense, at the time. Eddington's application of the basic laws of physics to the structure of the Sun and stars included a calculation of the temperature which must be maintained at the centre of the Sun in order to provide its visible surface luminosity and to produce a pressure strong enough to hold the star up against the inward pull of gravity. This temperature is about 15 million degrees, on the Kelvin scale. Temperature is simply a measure of the kinetic energy shared among a lot of particles – in the case of the air in my study, the temperature is an indicator of how fast, on average, a typical molecule in the air is moving. At the heart of the Sun, the temperature is a measure of how fast, on average, a typical proton is moving, as it bounces around, colliding repeatedly with its neighbours. The greater the temperature, the greater the speed of the protons, and the greater is the pressure they exert on each other and on the layers of the star above them – which, of course, is how Eddington was able to deduce the temperature in the first place. The problem was that the temperature he found, 15 million K, didn't seem to be enough for the fusion process to take place.

Fusion between two protons (to take the simplest case) will only happen if they collide with enough energy to overcome the electric repulsion forces between them. Going back to the volcano analogy, the two particles have to rush in from opposite sides of the hill with sufficient speed to carry them up the slopes and over the top, where they can drop down into the crater inside, the region where the strong nuclear force dominates. A sea of protons at a temperature of 15 million K, even at the density of the heart of the Sun, simply does not have enough energy for a large enough number of collisions to happen in the right way for fusion to occur and provide the heat of the Sun – or so it seemed when physicists first ran through the calculations. Either the physicists were wrong, or Eddington's calculations were wrong.

Eddington was sure of his own numbers, and is widely reported as having told his doubting colleagues to 'go and find a hotter place' (go to hell, in other words). But the solution to the puzzle was soon forthcoming, in large part thanks to the equations Gamow developed to describe how an alpha particle can get *out* of the potential well (the volcanic crater) of a massive nucleus. The key to an improved understanding of the fusion process turned out to be a phenomenon of quantum physics called uncertainty. In a nutshell, what it means in effect is that if two protons are pressed close together during a collision, they *don't* have to climb all the way over the potential barrier caused by the electric forces. If they get *near* to the top of the barrier, they can tunnel through and drop into the potential well inside, even if they don't *quite* have enough energy to get over the top. This is exactly the explanation Gamow found for how particles can *escape* from a nucleus without going 'over the hill', but running the other way around. The tunnel effect provided a means for a significant amount of fusion to occur even at 15 million K.

So things stood in the mid-1930s. It had become clear, at least to the experts, that the Sun must be powered by nuclear fusion and an associated conversion of mass into energy. Since the Sun is made up of 70 per cent hydrogen and 28 per cent helium, and since the helium nucleus is a particularly stable one, it also seemed clear that the key process must involve the conversion of hydrogen, four protons at a time, into helium. But nobody knew exactly how the protons were converted into helium nuclei, until Hans Bethe (the same Bethe later to achieve immortality as the 'absent' middle author of the alpha, beta, gamma paper) came along in 1939 and showed one way in which the trick could be carried out. His work started a new line in astrophysical studies, the investigation of how the elements are built up inside stars (stellar nucleosynthesis). It was taken up by a new generation of astrophysicists (after a brief delay caused by World War Two), and led to the award of Nobel Prizes to Bethe himself, in 1967, and to William Fowler in 1983.

Cycles and chains in stars

Hans Bethe was born in Strasbourg (then in Germany, now part of France) in 1906. He studied at the universities of Frankfurt and Munich, and after obtaining his PhD in 1928 he taught physics and did research in German universities until 1933. With the rise of the Nazis in Germany, Bethe moved to Britain, spending a year at the University of Manchester and a year in Bristol, where he worked until 1935. When he left Bristol University, Bethe moved to Cornell University in the United States, and became a naturalized American citizen. In the 1940s, Bethe worked on the Manhattan Project, the design and manufacture of the first atomic bombs; later he served as a delegate to the first International Test Ban Conference in Geneva, helping to achieve the agreement between the superpowers banning atmospheric testing of nuclear weapons, and he continued to advise on nuclear disarmament issues after the signing of the treaty that came out of those negotiations. But his major contribution to science came with the work carried out in 1938, and published in 1939, that showed how to generate energy by making helium inside a star. It was largely for this work that he received the Nobel Prize.[4]

The first mechanism Bethe came up with for energy generation inside stars involved the presence of heavier nuclei, particularly carbon, as well as hydrogen. He calculated that under the right conditions the collision of a proton with a nucleus of carbon-12 could produce a nucleus of nitrogen-13, which would then emit a positron and become a nucleus of carbon-13. In the same way, further collisions between the nucleus and other protons (hydrogen nuclei) at the heart of the star would produce first a nucleus of nitrogen-14, then one of oxygen-15 which decays by emission of a positron to nitrogen-15. At that point, when another proton comes along the most likely result of the collision is not to produce a nucleus of oxygen-16, but for four nucleons to split off as an alpha particle, forming a helium nucleus and leaving behind a nucleus of carbon-12, exactly what you started with. Only about once in a thousand collisions would a nucleus of oxygen-16 form, and even that, after a couple more protons had been taken on board, would emit an alpha particle and decay into nitrogen-14, rejoining the cycle. The net effect is that four protons have been converted into one helium nucleus, with the appropriate release of energy. And since the carbon-12 is put back where it came from to act as a catalyst for further fusion cycles, only a little carbon is needed to ensure a lot of nuclear fusion and the production of a lot of energy.

The process involves carbon, nitrogen and oxygen nuclei, as well as those of hydrogen and helium, and it goes round in a circle, starting and ending with carbon-12, and converting four protons into one helium nucleus along the way. Naturally, it is called the carbon nitrogen oxygen cycle, or CNO cycle.[5] Bethe and a colleague, Charles Critchfield, then came up with an alternative route for hydrogen fusion inside stars. This is the step-by-step process starting with hydrogen nuclei and building first deuterium, then helium-3 and helium-4, either directly or by collisions between helium nuclei that produce nuclei containing seven nucleons each, ready to be converted into two alpha particles by the addition of just one more proton. This process – the one we met in the previous chapter, that halts at element number four, helium – starts with two protons colliding to make deuterium; it is called the proton-proton (or pp) chain.

Bethe's calculations showed that such reactions could proceed under the conditions of temperature and pressure which Eddington had shown must exist inside stars. In fact, for our Sun the dominant process of energy production is thought to be the pp chain, which operates at the required efficiency at a temperature of about 15 million

$$C^{12} + H^1 \longrightarrow N^{13} + \gamma$$
$$N^{13} \longrightarrow C^{13} + \beta^+ + \nu$$
$$C^{13} + H^1 \longrightarrow N^{14} + \gamma$$
$$N^{14} + H^1 \longrightarrow O^{15} + \gamma$$
$$O^{15} \longrightarrow N^{15} + \beta^+ + \nu$$
$$N^{15} + H^1 \longrightarrow C^{12} + He^4$$

Or, once in 1,000 times:

$$N^{15} + H^1 \longrightarrow O^{16} + \gamma$$
$$O^{16} + H^1 \longrightarrow F^{17} + \gamma$$
$$F^{17} \longrightarrow O^{17} + \beta^+ + \nu$$
$$O^{17} + H^1 \longrightarrow N^{14} + He^4$$

Net effect

$$4H^1 \longrightarrow He^4$$

Table 6.1 The CNO cycle.

K; the CNO cycle works better at higher temperatures, above 20 million K, and so it is more important in more massive stars, which have to be hotter inside in order to be held up against the possibility of gravitational collapse. But these interesting details are not directly relevant to the story of the search for the Big Bang and the ultimate fate of the Universe.

Cooking the elements

Bethe's work did not explain how carbon came to be inside stars in the first place, but it did explain how a star like our Sun could make use of the energy which ought to be available from the conversion of hydrogen into helium. The calculations were based on measurements of the way particles interacted with one another – their cross sections – under laboratory conditions, and these cross sections then had to be extrapolated to the conditions which the laws of physics said must exist inside stars. This is a huge step, scientifically speaking – not so much because it turns out that the behaviour of particles

$$H^1 + H^1 \longrightarrow D^2 + \beta^+ + \nu$$
$$D^2 + H^1 \longrightarrow He^3 + \gamma$$
$$He^3 + He^3 \longrightarrow He^4 + 2H^1$$

---

or

$$He^3 + He^4 \longrightarrow Be^7 + \gamma$$
$$Be^7 + e^- \longrightarrow Li^7 + \nu + \gamma$$
$$Li^7 + H^1 \longrightarrow 2He^4$$

---

or

$$Be^7 + H^1 \longrightarrow B^8 + \gamma$$
$$B^8 \longrightarrow Be^8 + \beta^+ + \nu$$
$$Be^8 \longrightarrow 2He^4$$

Net effect

$$4H^1 \longrightarrow He^4$$

Table 6.2 The pp-chain.

measured in the lab *can* be extrapolated to tell us how stars work, but because of the way this discovery changes the whole conceptual approach of astrophysicists. Thanks to Eddington and Bethe, astrophysics became an *experimental* science; it was now possible to plan to carry out experiments here on Earth, with colliding beams of particles, that would unlock the secrets of nuclear fusion in the hearts of stars. When Bethe's first paper on the CN cycle (as it then was) appeared in the *Physical Review* (Volume 55, page 434) in 1939, it made just this impact on a group of nuclear physicists working at the Kellogg Radiation Laboratory, part of the California Institute of Technology.

There, the senior physicist Charles Lauritsen, and two younger men, his son Thomas and Willy Fowler, were involved in measurements of the cross sections for interactions involving carbon and nitrogen nuclei bombarded with a beam of protons. Bethe's paper showed them that they were studying in the laboratory processes which occur in the Sun and stars. Forty-four years later, in his Nobel address, Fowler said 'it made a lasting impression on us'.[6] So strong was the impact, indeed, that when the laboratory got back to basic nuclear research in 1946, the senior Lauritsen decided to concentrate on the study of just those nuclear reactions that were thought to take place inside stars. Fowler took up the challenge, and became the leader of the Caltech investigation of how stars work.

Fowler was an ebullient, extrovert character who remained active in research in the 1980s. 'I intend to remain active,' he told reporters when the award of his Nobel Prize was announced, 'until they carry me out.' He was born in 1911, in Pittsburgh, and studied physics at Ohio State University, graduating in 1933, before he moved to California and gained his PhD at Caltech in 1936. His main base remained there ever since, although he was often to be found spending a few months at other research centres around the world. Fowler, and the Kellogg lab, played a key role in developing our understanding of stellar nucleosynthesis, and also in the modern calculations of precisely how much helium could have been produced in the Big Bang. But, as Fowler also acknowledged in his Nobel address, the 'grand concept' of nucleosynthesis in stars came from Fred Hoyle (now Sir Fred) in two papers published in 1946 and 1954.

The dates are important, especially to Hoyle. When Fowler's Nobel award was announced, many of the popular accounts of the work for which it was given mentioned Hoyle's contribution, and some of them were quick to point out that the need to find a way to synthesize the elements inside stars was an obvious requirement for one of the inventors of the Steady State theory. After all, if there was no Big Bang, then the Steady Staters *had* to make their elements inside stars! Since the Steady State theory is now discredited, these reports suggested that Hoyle had hit on the right idea for the wrong reason – and hinted that, perhaps, this was why he hadn't shared the award with Fowler. But the Steady State theory did not 'come along', as Hoyle pointed out to me at the time of Fowler's award, until 1948, whereas his first paper on nucleosynthesis was written in 1945 and published in 1946.[7]

And besides, the Big Bangers can't make anything heavier than helium in the Big Bang, in any case.

Hoyle gives the outward appearance of being a prickly character whose reluctance to suffer fools gladly has not endeared him to the establishment, and led to his premature retirement from his posts at the University of Cambridge in 1973, at the age of fifty-eight. In fact, he is basically a shy introvert who lives for his work and has clearly developed a more forceful persona as a means of communicating the importance of his ideas to others. He was born in Bingley, in Yorkshire, in 1915, and sometimes gives his opponents the impression that his 'Yorkshire blunt speaking' is carefully cultivated – sometimes what might otherwise be regarded as downright rudeness seems to be forgivable as the traditional frankness of a Northerner. But he has clearly been deeply wounded by the failure of the British scientific establishment, in particular, to recognize fully the merits of his own work and ideas about how science ought to be run. If he were really as thick-skinned and insensitive as the image implies, he would surely have soldiered on longer in Cambridge.

His early career, though, followed conventional lines, through the local grammar school to Emmanuel College in Cambridge and then a Fellowship at St John's College in 1945. In 1958 he became Plumian Professor, and he was the inspiration behind, and first Director of, the Institute of Theoretical Astronomy, founded in Cambridge in 1967. But this was the peak of his establishment career; although Hoyle served on many high-level committees (while finding enough spare time to write popular books of science fact and science fiction), was elected a Fellow of the Royal Society and honoured with a knighthood, he fell out with the authorities in Cambridge concerning the role and development of astronomical research within the university, he disagreed violently with the administrators of British science about how science as a whole should be funded and carried out, and in the 1970s and 1980s he alienated many of his scientific colleagues by espousing ideas about the origin of life in the Universe which many regarded as unsound. But none of that, nor any opinions other astronomers may hold about the Steady State model or any other of Hoyle's wide-ranging views about the nature of the Universe, should be allowed to detract from the key role he played in the discovery of how elements are cooked inside stars.

The problem began to be a pressing one for astronomers not because of any interest in the Big Bang or Steady State models of the Universe, but because their improving observations of stellar spectra increasingly showed, in the 1940s and 1950s, that different stars contained different amounts of the different elements. You might speculate that the material from which stars formed came from a Big Bang, or you might speculate that the material to make new stars is being created continuously in the spaces between the galaxies. But when you find that there are systematic differences in the composition of the stars, with some stars richer in heavier elements than others, you begin to suspect that some of those elements are being manufactured out of the primordial material (whatever that may have been) inside the stars themselves.

Hoyle's 1946 paper presented, for the first time, a clear exposition of the basic ideas of nucleosynthesis within the accepted framework of stellar structure and evolution and using the best information about nuclear reaction rates, cross sections and so on. As Gamow's team struggled to find a way to make elements heavier than helium in the Big Bang, Hoyle struggled to find ways to make them inside stars, visiting Caltech for the first time in 1953 and soon joining forces with Fowler. The key problem was how to get past the instability of the boron-8 nucleus. The only way to do this was to invoke a triple collision, with three alpha particles colliding almost simultaneously to form a nucleus of carbon-12. Gamow couldn't make this work in the Big Bang, because even a few minutes after the moment of creation matter in the Universe was spread too thin, and the temperature was too low, for such collisions to happen often enough to produce the amount of heavy elements we see in the Universe today. Inside stars, however, it is both hot and dense, and the inside of a star stays hot and dense for many millions of years, giving a much better opportunity for even relatively rare triple collisions to occur often enough to produce a significant amount of carbon.

The idea looked good, but it ran into problems rather like the problems Eddington encountered when the physicists told him the Sun was not hot enough for hydrogen fusion to occur. When Ed

Salpeter, a physicist visiting the Kellogg lab in 1951, made the appropriate calculations he found that the cross sections still weren't big enough. You could make *some* carbon-12 inside stars, but nowhere near enough. Now Hoyle made a dramatic contribution. He came to Caltech in 1953 convinced that all the heavy elements are made in stars. He had tackled the calculation from the other end, using the observed measurements (from spectroscopy) of the abundances of heavy elements in stars to deduce how fast the triple collision reaction *must* proceed, and he found that it had to go much faster than Salpeter had calculated. So he predicted that the carbon-12 nucleus must be capable of existing in what is called an excited state, a state which has more energy than the minimum value appropriate for that nucleus. If such an excited state existed, with precisely the right amount of energy, then, and only then, could the collision of three alpha particles together be encouraged to form carbon-12 nuclei sufficiently often to make all the carbon observed in the spectra of stars. The reaction is encouraged by a process called resonance between the energy state of the three alpha particles and the energy state of the carbon-12 nucleus, and the resonance only occurs if the carbon-12 energy level is just right, which is how Hoyle was able to predict what the excited energy state of carbon-12 must be.

Hoyle badgered the physicists at Caltech until a group of them went away to look for an excited state of carbon-12, using a reaction involving deuterium particles colliding with nitrogen-14 nuclei to make carbon-12 plus an alpha particle. And they found it, almost exactly where Hoyle had predicted.

This was still one step short of proving that the excited state of carbon-12 could be produced by the interaction of three alpha particles, but now Fowler, working with the two Lauritsens and Charles Cook, manufactured 'excited' carbon-12 from the decay of boron-12. They found that although some of the excited carbon-12 then fell back into its minimum energy state (the ground state) and stayed as carbon-12, some of it broke up into three alpha particles. Other things being equal (and in this case they are) these kinds of nuclear reaction are reversible. Since excited carbon-12 can decay into three alpha particles, there is no doubt that three alpha particles can combine to form excited carbon-12. Here was the proof that as well as burning hydrogen to make helium, stars could burn helium to make carbon. The helium-burning process explained how large stars, called red giants, are kept hot, and it also got astrophysicists over the hurdle of nucleosynthesis at element number eight. And, of course, it provided the carbon needed for the CNO cycle to operate. By looking at information from stellar spectroscopy, Hoyle had correctly predicted what physicists would find in experiments in the lab here on Earth. That gave them the confidence to continue their measurements of interaction rates in the lab, and use the information to calculate the whole chain of reactions needed to build up all of the naturally occurring elements, in all their varieties or isotopes, in the stars.

In a brief survey of how stellar nucleosynthesis became understood, it is inevitable that the rest, after the production of carbon-12, may seem almost an anticlimax, just the icing on the cake that Hoyle, together with Fowler and his colleagues, had already baked. In outline, it was now easy to understand how all the elements are made in stars. Alpha particles (helium nuclei) are added to the nuclei, increasing their mass four units at a time; decays which eject electrons, positrons or neutrons then form the other elements and isotopes. For very heavy elements (heavier than iron), capture of individual neutrons, increasing the mass of the nucleus by one unit at a time, also becomes important. But there are no more yawning chasms to worry about like the one facing astrophysicists trying to build carbon-12 out of alpha particles. The rest was a question of painstaking, detailed work to measure all the necessary cross sections and reaction rates, and to fit these to the calculated conditions of temperature and pressure inside stars, and to the observed abundances of the elements deduced from stellar spectroscopy.

The cross-section measurements were no small task in themselves; extrapolating these from the relatively high energy conditions of collisions in accelerators in the lab to the much lower energy conditions of collisions between particles inside the stars required great skill; and the observers were stretched to find out just what the composition of the Universe is, anyway. But everything came together in the mid-1950s. Fowler spent the academic year 1954–5 in Cambridge, working with Hoyle and with Margaret and Geoffrey Burbidge, a British husband-and-wife team of astronomers. The collaboration continued long after

Fowler returned to Caltech, with now Hoyle and then the Burbidges (sometimes all three) visiting the Kellogg lab in turn. In 1956, astronomers Hans Suess and Harold Urey published the best data yet on the relative abundances of all the naturally occurring elements; that same year, the four collaborators produced a short paper in *Science* on the origin of the elements, and in 1957 they followed it with a paper in *Reviews of Modern Physics*[8] which remains one of the classic scientific papers of all time. In alphabetical order, the paper was signed 'Burbidge, Burbidge, Fowler and Hoyle'; it is known to all astronomers simply as 'B^2FH', and no further reference is needed when citing it. The paper describes how all the naturally occurring varieties of nuclei, except for hydrogen and helium, are built up inside the stars – and in the words of the Swedish Academy of Sciences' announcement of Fowler's Nobel prize, this paper 'is still the basis of our knowledge in this field, and the most recent progress in nuclear physics and space research has further confirmed its correctness'. On a more personal note, I still recall the thrill I got when I first came across the paper as a graduate student in 1966; the awe of knowing that the equations in the paper I held explained where all the atoms in my own body (except for the primordial hydrogen) came from, and how all those atoms had been cooked in stars. As far as any piece of scientific investigation could, this paper closed a chapter of research, with the complete answer to a major puzzle not just in science but in philosophy, an answer which it had taken Hoyle and his collaborators just ten years to track down, from the time he had published his first landmark paper on the subject.

The B^2FH paper marked the end of the puzzle of stellar nucleosynthesis, except for dotting the i's and crossing the t's. But it didn't mark the end of the puzzle of all nucleosynthesis. Starting out with about 70 to 75 per cent hydrogen and 25 to 30 per cent helium in the first generation of stars in the Universe, the astrophysicists could show how all the heavier elements were made, and they could guess that these heavy elements are then scattered across space when some old stars explode, as novae and supernovae, to enrich the mixture of gases from which later stars are made. Our own Sun is relatively young; it contains the recycled material from older stars, long since dead, and that is where its 2 per cent of heavy elements, and the material of which the Earth and ourselves are made, comes from. Sixty-five per cent of your body weight is oxygen; 18 per cent is carbon. It has all been through the triple alpha capture process, among other reactions, inside stars. But stellar nucleosynthesis could not explain why there was so much helium in the Universe. This 'helium problem' bedevilled astrophysicists in the late 1950s and early 1960s, in spite of Gamow and Alpher's earlier success in making helium in the Big Bang. Perhaps because they had failed in their intended objective of making *all* the elements in this way, their success with helium was overlooked; whatever the reason, it was left to Fowler, with his student Robert Wagoner and, once again, Fred Hoyle,[9] to apply their improved knowledge of nuclear reaction rates to the conditions that were thought to have applied in the first few minutes of the creation of the Universe.

With reaction rates for almost a hundred nuclear processes determined by Fowler's group included in the calculations, and using the fact that other reactions had been shown in the laboratory to be insignificant, this team established that no significant quantities of any element heavier than helium could be produced in the Big Bang, that the proportion of helium-4 coming out of the Big Bang ought to be about 25 per cent, and that associated with it there should be deuterium, helium-3 and a trace of lithium-7 in proportions similar to those in which these elements are found in the Solar System.

Wagoner, Fowler and Hoyle published their findings in 1967. It was another personal landmark for me – my first visit to Cambridge was to hear the team give a report of these findings, and I clearly recall the penetrating questions asked at the gathering by an unknown Cambridge research student, Stephen Hawking. That visit, and the excitement of the meeting, were instrumental in encouraging me to move from the University of Sussex to Cambridge, intending to do research in cosmology. I ended up working on problems in stellar structure,[10] but I still think the trip was worthwhile! Far more importantly, though, the Wagoner, Fowler and Hoyle collaboration had a major impact on cosmology. 'It was this paper,' says Sir William McCrea, 'that caused many physicists to accept hot Big Bang cosmology as serious quantitative science.'[11]

The science was so 'serious', indeed, that it was able to address a much more subtle problem than the abundance of helium in the

Universe. The helium is manufactured by the fusion of deuterons, nuclei of heavy hydrogen (deuterium). Very nearly all the deuterons are used up in the process, but a small fraction – between 0.01 and 0.001 per cent – of all the hydrogen in the Universe today seems to be in the form of deuterium, judging by spectroscopic studies of stars and galaxies. Although the proportion of helium that emerges from the Big Bang is not very sensitive to such factors as the overall density of the matter emerging from the Big Bang, it turns out that the deuterium abundance *is* a very sensitive indicator of density. If the model used in the calculations is more dense, then the reactions proceed faster and the deuterons are used up more quickly. And the kind of calculation carried out by Wagoner, Fowler and Hoyle indicates that the density of everyday matter in the Universe is less than the critical amount needed to make the Universe closed and ensure that it will one day collapse back into a fireball. This is the strongest piece of evidence in favour of the possibility that our Universe is open, and will expand forever. But it is not the last word on the subject. The latest ideas in cosmology, discussed later in this book, include the possibility that there may be a great deal of other matter in the Universe, in forms which did not take part in the nuclear reactions described by Wagoner, Fowler and Hoyle and their successors, and which might 'close' the Universe gravitationally, regardless of what the deuterium is telling us about the amount of ordinary matter – by which I mean atoms and atomic nuclei – that came out of the Big Bang. But that is getting ahead of the story as it stood in the late 1960s.

The investigation of nucleosynthesis had by then provided a beautifully dovetailing pair of pieces of evidence in favour of the hot Big Bang. All the investigations, from Gamow onward, had shown that no elements heavier than helium could be made in the Big Bang. They had to be made somewhere else. The equations of stellar nucleosynthesis showed that everything heavier than helium could indeed be made in stars, but that the observed amount of helium in the Universe could not; it must have been made somewhere else. Big Bang theory needed stellar nucleosynthesis; stellar nucleosynthesis needed the Big Bang. Together, the combination of a hot Big Bang and nucleosynthesis inside the stars provided a beautiful, complete picture of where *everything* came from.

And there was another piece of 'quantitative science' in that 1967 paper which helped to make the physicists sit up and take notice of the Big Bang theory. Wagoner, Fowler and Hoyle included in their work the first quantitative application of a new discovery, the cosmic microwave background radiation, in establishing the parameters of their model. This cosmic background is the radiation that was half-predicted by Gamow and his colleagues in the 1950s, and then forgotten about. It is the second piece of powerful evidence that the Universe did indeed emerge from a hot fireball – the second key to the Universe.

The lost years

The idea of taking the temperature of the Universe and using the measurement to find out more about the Big Bang in which the Universe was born may have been too farfetched for physicists and astronomers to take seriously in the 1950s. But that doesn't mean the idea was totally ignored, and more than one astronomer has looked back ruefully to that decade and metaphorically kicked himself for failing to follow the idea through to its logical conclusion. Indeed, Gamow, Alpher and Herman must have shared these feelings – not least since some astronomical observations which clearly implied a background temperature of the Universe of about 3 K had already been carried out in the 1930s, and were certainly known to Gamow and his colleagues in the 1950s.

These observations, like so much of our information from space, depend on spectroscopy. In the 1930s, astronomers began to identify, for the first time, spectral features corresponding to the presence of molecules in interstellar space. Starlight carries with it the spectral signature of the atoms (or more accurately ions, atoms with some electrons stripped off) present in the atmosphere of the star. The characteristic lines stand out as either bright emission lines (radiating energy) or dark absorption lines (absorbing energy from the star below) in the electromagnetic spectrum. The strength of these lines, and the extent of the ionization they reveal, enables astronomers to deduce the temperature of a distant star, as well as to determine its composition. But there are also lines in some spectra which correspond to compounds that could not possibly be stable at the temperature at the surface of a star. One of the earliest of these to be identified is cyanogen, CN, a stable pairing of one carbon atom and one nitrogen atom to produce what is known as a radical. Such compounds occur not in the stars themselves, where the heat would soon break them into their component atoms, but in cool clouds of gas and dust between the stars. Their presence is revealed by the dark lines they impose on the light from distant stars shining through the cool clouds.

Just as observations of stellar spectra reveal the temperatures of the stars, so observations of these absorption spectra can reveal the temperature of the clouds of interstellar material. In 1940, W. S. Adams, at Mount Wilson, observed interstellar spectral lines corresponding to an energetic state of cyanogen, and Andrew McKellar, of the Dominion Astrophysical Observatory in Canada, interpreted those observations as indicating a temperature for the interstellar clouds of about 2.3 K. By 1950, the result was enshrined in a standard textbook of spectroscopy,[12] and was very well known to astronomers, including Gamow. But nobody thought of interpreting the temperature of the coldest clouds of material found in space as 'the temperature of the Universe'. One of the nearest misses came in 1956, when Fred Hoyle and George Gamow were cruising around La Jolla, in southern California, in a brand-new white Cadillac convertible.

Hoyle recounted the tale in an article in *New Scientist* published in 1981.[13] He was visiting Willy Fowler and his colleagues at Caltech that summer, and Gamow called them from La Jolla to invite Fowler, Hoyle and the Burbidges down for a visit. Gamow was in La Jolla because he was spending two months as a consultant with General Dynamics, a job which was very lucrative (the two months' consultation fee paid for the white Cadillac convertible) and apparently required little real work, but for which Gamow was obliged to stay in La Jolla somewhere (even on the beach) to be on call immediately if his services were required. So the B^2FH team made their way, not too reluctantly, south to La Jolla. At that time, Gamow's estimates for the temperature of the Universe today were in the range from about 5 K to a few tens of K; Hoyle, as a Steady Stater, thought there should be no background radiation at all. So they both missed the truth that lay under their noses. Hoyle takes up the story:

There were times when George and I would go off for a discussion by ourselves. I recall George driving me around in the white Cadillac, explaining his conviction that the Universe must have a microwave background, and I recall my telling George that it was impossible for the Universe to have a microwave background with a temperature as high as he was claiming, because observations of the CH and CN radicals by Andrew McKellar had set an upper limit of 3 K for any such background. Whether it was the too-great comfort of the Cadillac, or because George wanted a temperature higher than 3 K whereas I wanted a temperature of zero K, we missed the chance . . . For my sins, I missed it again in exactly the same way in a discussion with Bob Dicke at the twentieth Varenna summer school on relativity in 1961. In respect of the microwave background, I was evidently not 'discovery prone' . . .

The Bob Dicke that Hoyle discussed the problem with in Varenna in 1961 deserves a special place in the hall of missed opportunities, for, quite apart from that conversation, he missed the chance to go down in history as 'the man who took the temperature of the Universe' not once but *twice* – and on the second occasion he had forgotten about his own earlier work on the problem! Just a year younger than Hoyle, Dicke was born in St Louis, Missouri, in 1916. He graduated from Princeton in the late 1930s, completed a PhD at Rochester University in 1941, and worked on radar at MIT during the war, before joining the faculty at Princeton in 1946. He has stayed there ever since, to become Chairman of the Department of Physics and Albert Einstein Professor of Science. Dicke is nobody's fool. But he, too, could not see in the 1940s what now seems obvious with hindsight.

During his time at MIT, Dicke developed an instrument for measuring very short wavelength radio radiation, in the microwave part of the electromagnetic spectrum. The instrument is called a Dicke radiometer; its principles are incorporated in modern instruments designed to do the same job. With three of his colleagues, Dicke pointed one of these instruments at the sky, looking to see if there was any background of microwave radiation from the external galaxies. One way of interpreting the strength of such radiation is in terms of temperature; Dicke and his colleagues concluded that there was a background radiation with a temperature below 20 K, the limit that could be set by their instrument, and they wrote a paper reporting this result. It was published in the *Physical Review*, in the same volume in which Gamow's 1946 paper on nucleosynthesis appeared. The paper by Dicke's team appeared first (Volume 70, page 340); Gamow's came along a little later (Volume 70, page 572). There is nothing to link the two papers, but they appear in the same bound volume of the journal, and every student, or more senior researcher, in the 1950s who went to look up Gamow's paper, perhaps following the story back from the alpha, beta, gamma paper, or from the work by Alpher and Herman, literally held in his (or her) hands the evidence that the cosmic fireball really had existed. Anyone – Hoyle, Gamow, or some unknown student – looking up Gamow's paper might have come across the Dicke team's paper and put two and two together; but it didn't turn out that way. Sometimes, scientific discoveries seem to have a will of their own, waiting until the time is ripe for them to happen.

By the early 1960s, Dicke himself had forgotten all about this measurement. But his thoughts were turning to cosmology, and in surprising, but seemingly complete, ignorance of the pioneering efforts of Gamow, Alpher and Herman, he independently investigated the implications of a model universe which collapses down from a very great size into a fireball, then bounces away from the very high density state and expands. Dicke was intrigued by the idea that the Universe might be in the expansion phase of an oscillation which could continue forever, with each cycle of expansion followed by one of collapse, each collapse followed by a bounce and a new phase of expansion. And he needed the collapse to continue down to a state of very high temperature and density before the 'bounce' occurred, so that all the material in the collapsing Universe would be broken back down into neutrons and protons before a new phase of expansion began – there must be no 'information', as it were, carried over from one cycle of the Universe to the next, and to anyone living in the expanding Universe it would be just as if the Universe had been created in a Big Bang.

All this was still very much in the spirit of cosmology as a game, an intellectual exercise. But Dicke's experience as an observer (albeit half-forgotten experience) set him and his colleagues on the right trail at last. He gave a young researcher at Princeton, P. J. E. Peebles, the task of working out the way the temperature of such a model universe would change as it evolved; repeating, unknowingly, the calculations Alpher and Herman had carried through more than fifteen years before, Peebles found that if the Universe we live in had started out in a hot Big Bang, then it should be filled with a background sea of radiation with a temperature of about 10 K. In 1964, in the light of Peebles's calculations, Dicke encouraged two other members of the Princeton research staff to carry out a search for this radiation. P. G. Roll and D. T. Wilkinson set up a detector (a version of the Dicke radiometer), and they began to construct a small antenna on the roof of the physics lab at Princeton in order to detect any cosmic background radiation with a temperature of a few K. Then, on the point of making an epochal discovery, the rug was pulled from under the Princeton team. Dicke received a phone call from a young man at the Bell Research Laboratories, just 30 miles away from Princeton, at Holmdel, in New Jersey. The caller, Arno Penzias, and a colleague, Robert Wilson, had been getting some funny results from their radio telescope, a 20-foot horn antenna used in some of the early experiments with communications satellites. Someone had suggested Dicke might be able to explain this puzzling cosmic background radiation; perhaps they could all get together to discuss it . . .

The echo of creation

Arno Penzias comes from a Jewish family in Munich. He was born in 1933, on the same day (26 April) that the Gestapo was formed. The family was one of the last to get out of Nazi Germany to England in 1939 – Arno and his brother were sent on in the spring, and were later followed by his father and, finally, mother. Reunited, the family left England in December 1939, sailing for New York, where they landed in January 1940 and stayed. Education provided the opportunity for this son of an impoverished immigrant family to make his way in the world, and in 1954 Penzias graduated from the City College of New York with a degree in physics. After two years in the Army Signal Corps, he joined Columbia University as a graduate student, working for his PhD, which was awarded in 1962.

Penzias worked at Columbia with Charles Townes, a physicist who played a key role in the development of masers and lasers,[14] and who was to receive a Nobel Prize for his efforts in 1964. A maser can be used as the basis of an amplifying system to detect weak radio emissions, and Penzias built a maser receiver designed to operate at 21-centimetre wavelength. This is the natural wavelength at which hydrogen gas radiates, and Penzias hoped to detect the 21-centimetre 'line' of intergalactic hydrogen. But he failed, largely because there isn't any intergalactic hydrogen to be observed, and he was only able to set an upper limit on how much hydrogen there might be between the galaxies. This kind of 'failure' is not at all uncommon for a doctoral student, or even for other research projects, and clearly the examiners at Columbia were happy that Penzias had carried out his work effectively, even if it came up with a negative result. But he took a harsher view of his own work. As he told Jeremy Bernstein, author of the book *Three Degrees Above Zero*, 'I just got through Columbia by the skin of my teeth . . . it was a dreadful thesis.' Dreadful thesis or not, though, the association with Townes and his first attempt at radio astronomy profoundly influenced the rest of Penzias' career.

Townes had come to Columbia University from Bell Labs in 1948. Bell Labs was originally a research division of the Bell Telephone Company, later swallowed up by AT&T, and more recently involved in antitrust legislation to break up the conglomerate, which has meant some uncertainty about the continuing role of Bell Labs as an independent research institution. Whatever their fate, though, the Bell Labs have a proud history of research, including the first discovery of radio noise from space by Karl Jansky, a Bell researcher and the founder of radio astronomy, in the 1930s. Through Townes's continuing contacts with Bell Labs, Penzias joined their Radio Research Laboratory at Crawford Hill near Holmdel in 1961, shortly before the award of his PhD. Although fundamentally devoted to practical research to benefit the parent company, Bell Labs has always maintained a tradition of academic research as well, which helps to ensure that first-rate scientists are always eager to join the team, and which keeps the practical side of Bell Labs in touch with new developments coming out of the universities and other academic institutions. Penzias worked at first on problems associated with the systems that were used for the first satellite communications links, using the Echo and Telstar satellites. Then he was allowed to turn to radio astronomy again, making only a little headway before he was joined at Crawford Hill by another would-be radio astronomer, Robert Wilson. One radio astronomy post was the ration allowed, so Penzias and Wilson split it between them, each devoting half their time to radio astronomy, and half to other projects.

Wilson comes from a very different background to Penzias. Born in Houston, Texas, in 1936, he is the son of a chemical engineer, and both his parents went to college. With straight A's in all his science courses at Rice University, in Houston, when Wilson graduated in 1957 he was offered places in the graduate schools at both MIT and Caltech, the two premier scientific research institutes in the US; he chose Caltech, but with no clear idea of just what line of research he would like to take up. There, he was influenced by two British astronomers – Fred Hoyle, who taught the cosmology course during a spell as visiting professor at Caltech, and whose presentation left Wilson with a fondness for the Steady State theory; and David Dewhirst, who suggested that Wilson might like to work with John Bolton, an Australian radio astronomer then at Caltech. So Wilson worked with Bolton on a radio survey of the Milky Way, mapping out the clouds of hydrogen gas in our own Galaxy. The result wasn't exactly world-shatteringly important; the map was made, and confirmed the accuracy of a similar map made by a group in Australia. Wilson remains as deprecating as Penzias about the quality of his first research project: 'Frankly, I don't think much scientifically ever came out of the thesis, although it was a good learning experience, and I did get a chance to meet most of the world's radio astronomers, who came through Caltech to visit.' (Bernstein, *Three Degrees Above Zero*.) Whereas Penzias left Columbia just before completing his PhD, Wilson stayed on at Caltech for a year after completing his, in 1962. So it was in 1963 that, hearing about Bell Labs' interest in radio astronomy and the availability of the still relatively new horn antenna at Crawford Hill, he decided to take the plunge, and joined Penzias in New Jersey.

The antenna had been put up to work with the Echo series of satellites. These were simply large metalled balloons that inflated in orbit, and were used to bounce radio signals around the world. They had no amplifiers of their own, but acted like mirrors in the sky, so the signals were pretty weak by the time they got back to the ground stations, and needed to be caught by a good antenna system and amplified considerably if they were to be any use. With the advent of active communications satellites – Telstar and its successors, which amplify the signals they receive from the ground before they rebroadcast them to other ground stations – the designed role of the Crawford Hill antenna was at an end, so Penzias and Wilson were allowed to take the communications receiver out and turn the antenna into a radio telescope. This took several months. They wanted the new receiver to be as sensitive as possible, so that it could detect very weak astronomical radio noise. So they had to eliminate, as far as they could, all the sources of noise in the electrical systems used to amplify the radio waves from space. This noise is a bit like the static you get on an AM radio; some of the hiss of background noise is from stray radio waves (including radio waves from space), but some is just due to the inefficiency of the radio receiver itself. The static, or background noise, can be measured in terms of temperature, and the engineers working with the Crawford Hill antenna on tests with the Echo satellites had noticed that there was a little more static than they could explain in their system. In effect, the antenna was too hot; in an article which appeared in the *Bell Systems Technical Journal* in 1961, one of the engineers, E. A. Ohm, reported an excess noise, after subtracting out everything that could be explained away, equivalent to radiation with a temperature of about 3 K. This wasn't enough to disrupt the Echo communications system, so the engineers weren't too worried about it. But it was just the sort of thing that Penzias and Wilson had to track down and eliminate, or at least identify, before they could begin their planned program of radio astronomy research.

While Penzias and Wilson were trying to track down this infuriating source of noise in their system – even going so far as to clean out pigeon droppings from the horn itself, with no effect – the Princeton team was calmly proceeding with the plans to construct an instrument to detect the cosmic background radiation. At the same time, in 1964, over in England Fred Hoyle (that man again!) and Roger Tayler were beginning to move along the same lines, with calculations of the background temperature of a Big Bang Universe today. And in the

Soviet Union there was a veritable flurry of activity. Ya. B. Zel'dovich had also carried out the calculation which showed that in order to explain the observed abundances of hydrogen, helium and deuterium in the Universe it must have started in a hot Big Bang and have a temperature of a few K today; he even knew of Ohm's article in the *Bell Systems Technical Journal*, but misunderstood Ohm's terminology and thought that his measurements implied that the background temperature of the Universe was less than 1 K. Another Soviet researcher, Yu. N. Smirnov, calculated a background, or relict, radiation temperature in the range of 1 to 30 K and, jumping off from Smirnov's calculations, A. G. Doroshkevich and I. D. Novikov wrote a paper discussing the implications of various existing radio astronomy measurements in terms of the microwave background. They concluded that the best antenna then existing in the world for a search for this radiation was the Bell Labs antenna on Crawford Hill, and they suggested in their paper that the antenna be used for this purpose. All of this work was being carried out, and most of it published, in 1964. The idea of the cosmic microwave background had clearly decided the time was ripe for it to come out into the open. But with all the interest in at least four research centres spread across two continents, Penzias and Wilson themselves remained blissfully ignorant of the solution to their puzzle of where the excess noise in their system was coming from.

The accounts of how that ignorance was broken differ slightly, but the essentials are the same. According to one version, Penzias had been to an astronomical gathering in Montreal, and was returning, in December 1964, in an airplane where he sat next to Bernard Burke, who was based at MIT. During the flight, he mentioned the problems he and Wilson were having eliminating the background noise from their system, and as a result Burke telephoned Penzias a few days later to put him on the trail of the Princeton group. The other version of the story has it that Penzias just happened to mention the background noise in a phone call to Burke that he initiated, to discuss other matters. Either way, there is no doubt that it was during a telephone conversation in January 1965 that Burke, at MIT, told Penzias, at Crawford Hill, that yet another astronomer, Ken Turner, of the Carnegie Institution in Washington, DC, had heard a talk by P. J. E.

Peebles, the Princeton theorist, in which he predicted a background noise of electromagnetic radiation filling the Universe, with a temperature equivalent of about 10 K. Burke suggested that Penzias get in touch with the Princeton group; Penzias phoned Dicke, and very soon all four members of the Princeton team made the half-hour drive to Crawford Hill to find out what was going on. At last the theory and observations had been put together; two plus two really *did* make four.

The Princeton team was much more excited about the discovery than Penzias and Wilson were. To the Princeton researchers, the observation was in line with a prediction made by theory (what they thought was their theory), a good example of the scientific method at work. To Penzias and Wilson, although it was a relief to have some explanation of the radio noise they were measuring, it still seemed that other explanations might come along. Besides, Wilson was reluctant to accept that the Steady State hypothesis was dead until more evidence came in. In particular, the measurements had only been made at one wavelength, just over 7 centimetres; they would have to be made at many other wavelengths, using different receivers, before the true nature of the background radiation could be reliably understood.

So the news, although it spread rapidly throughout the scientific community, appeared in print in an extremely modest way. Penzias and Wilson agreed with the Princeton team that each group should submit a paper to the *Astrophysical Journal*, to be published alongside each other. The Princeton paper was much the more exciting and interesting of the two, and appeared first (Volume 142, page 414); Penzias and Wilson's paper followed it, under the inauspicious title 'A Measurement of Excess Antenna Temperature at 4,080 Mc/s' (Volume 142, page 419). News of the discovery for which they were to receive a Nobel Prize in 1978 was put in context only by the sentence 'a possible explanation for the observed excess noise temperature is the one given by Dicke, Peebles, Roll, and Wilkinson in a companion letter in this issue.' But perhaps the most remarkable feature of that issue of the *Astrophysical Journal* is that *neither* paper makes any reference to the work of Gamow, Alpher and Herman. The omission was soon corrected, and later publications invariably gave credit to those pioneers, but not before they had all been deeply upset at the way their work had been ignored.

Later measurements at different wavelengths established beyond doubt that the 'excess noise' referred to by Penzias and Wilson is indeed a cosmic background of electromagnetic radiation, exactly the kind of 'black body' radiation, with a temperature close to 2.7 K, required by the Big Bang model of the origin of our Universe.[15] It really is the echo of creation, a leftover piece of the Big Bang that we are able to reach out and touch with our instruments. The discovery ranks with the most important scientific discoveries ever made, and it changed the face of cosmology by making the participants realize that they were not playing some intellectual game, but were dealing with equations that really could describe the origin of our Universe and everything in it. The question 'Where do we come from?' moved from the realms of philosophy into the realm of science with the recognition of the relict radiation for what it was. And this is why Gamow and his colleagues were ahead of their time – because they were almost alone, in the 1940s and 1950s, in *believing* the equations. Steven Weinberg, one of the physicists who turned to cosmology once the realization that cosmology was indeed a science spread with news of the background radiation, has summed the situation up appositely:

Gamow, Alpher, and Herman deserve tremendous credit above all for being willing to take the early universe seriously, for working out what known physical laws have to say about the first three minutes. Yet even they did not take the final step, to convince radio astronomers that they ought to look for a microwave radiation background. The most important thing accomplished by the ultimate discovery of the 3 K radiation background in 1965 was to force us all to take seriously the idea that there *was* an early universe.[16]

Lemaître heard the news shortly before he died in 1966. Gamow outlived him only by a couple of years. Had they lived a little longer, or had the discovery of the background radiation been made a little sooner, they might well have shared a Nobel Prize for developing the concept of the Big Bang, a concept made real by that discovery. But Nobel Prizes are never awarded posthumously, and when the Nobel Committee decided, in 1978, that the time had come to take note of the reality of the early Universe, they were faced with what must have seemed a ticklish problem – who to give the award to. There was no shortage of candidates. On the one hand, there was a pair of young radio astronomers who had found something funny but had no idea what it was until somebody else told them, and who even then didn't really believe it at first. On the other, there was a team which had between them predicted the existence of a background, built an instrument to detect it and, only a little after the fateful meeting on Crawford Hill, had found it just as predicted using their own instrument. Leaving aside all the near misses from the Russians, Hoyle and Tayler and so on, there was, and still is, a third 'hand' to be considered, Alpher and Herman, the surviving members of the Gamow team, who said it all first, even though they were ignored.

The award went to Penzias and Wilson. In the circumstances, it could hardly have gone anywhere else, without being spread so thin as to be ridiculous. Or could it? I wonder if the Committee entertained, even for one moment, what would have been an inspired decision. Why, after all, could the award not have gone to the person who first reported the detection of the 3 K background – E. A. Ohm? He may not have known what he had found, but then, neither did Penzias and Wilson, and Ohm *did* find it first.

Such speculation is idle, however. What is done is done, and cannot be undone. The same seems to be true of the Universe. It did start in a Big Bang, and has evolved steadily ever since. With that one measurement, the temperature of the Universe today, available to calibrate the Big Bang, cosmologists were able to refine their calculations and come up with what is now the standard model of creation, the story of the Universe from a fraction of a second after the beginning of time itself.

Chapter 7

The Standard Model

We live in an expanding Universe that is uniformly filled with a sea of very weak electromagnetic radiation, and which contains matter in clumps dotted uniformly (on the large scale) throughout its volume. In order to find out what conditions were like long ago, we have to imagine winding the clock back, so that the resulting model of the universe contracts. The effect of contraction is to increase the density of the model universe. The density of matter increases, because the same amount of matter is being squeezed into a smaller volume as time goes by, and the density of radiation increases as well. The increase in radiation density shows up as blueshift, a shortening of the wavelength of the radiation, and can also be expressed in terms of temperature, starting out from the present-day 3 K and getting hotter the further back in time we go.

From now on, I shall be describing the standard model of the Universe, the current 'best buy' among cosmologies, as if it were a description of the real Universe. This is necessary licence; it would break up the flow of the story too much to keep putting in cautions about how this is the best description of the Universe that we have, but that new developments may supersede it. If a model is properly worked out and expressed in equations, we can say what *must* happen in the model as it evolves. The main features of what *must* happen in the standard model look very much like what we see going on in the Universe around us, so the standard model is a good one. We hope that some of the things that must happen in the standard model, but which we cannot see directly in our Universe, also tell us what the real Universe is like, in places or times where we cannot observe it directly. But the model can never tell us what *must* happen in the real Universe. I will tell the story as if it were the history of our Universe; that is the only way to tell it coherently. But keep at the back of your mind the understanding that this story is really all about a mathematical model universe, one which bears such striking similarities to our Universe today that we think we can use it to get an understanding of what happened long ago when the Universe was young.

Looking back in time corresponds to compression of the model universe, as its present expansion is time-reversed. If this compression of the model universe continued for long enough, then, the laws of physics tell us, we would reach a point when the density of both matter and radiation became infinite. Clearly, the laws of physics that we know from experiments here on Earth, and from observations of the Universe as we see it today, are inadequate to describe infinite densities of matter and energy, and must break down at some stage in this imaginary journey back to the moment of creation. But if we leave aside, for the moment, the puzzle of exactly what happened in the first split second of creation itself, our observations of the expanding Universe are entirely adequate to tell us that the creation must have occurred between 10 and 20 billion years ago. For the sake of argument, we can pick the middle of the range of age estimates, and say that there was a time $t = 0$, 15 billion years ago, when the Universe came into existence in a state of extremely high density and very high temperature, and we can describe the evolution of the Universe up to the present day in terms of the time that has elapsed since $t = 0$. We do so by winding the clock back, in our imagination, from the present to as near as we can get to the state of infinite energy density, and then imagine letting the clock run forward again as the Universe evolves. So, to start, we need to check what we know about the present-day Universe.[1]

First, we know that it expands. Secondly, we know that something like 25 per cent of the material in the stars is helium, and that most of the rest is hydrogen. Thirdly, we know that the Universe is filled with radiation at a temperature of 3 K. In terms of photons, the particles which represent electromagnetic radiation, that means that there are about 1,000 photons in every cubic centimetre of the Universe, and this is about a billion times more than the total number of protons and neutrons in the Universe, assuming that, as observations suggest, the density of everyday matter is roughly that required for a flat universe, somewhere near to the boundary between the open and closed states. The energy of this radiation is about 1/4,000 of the energy of the visible matter in the Universe, given by $E = mc^2$. Matter dominates the Universe today – but that was not always the case.

When we wind the clock back, we can calculate, from the simple laws of physics, how the temperature of the radiation increases as the Universe contracts. The amount of energy locked up in each proton or neutron stays the same, but the amount of energy locked up in each photon increases as the radiation is squeezed and blueshifted. When the temperature was about 4,000 K, the energy in each photon was one billionth of the energy in each proton or neutron, and since there are a billion times more photons in the Universe, the total energy in the radiation matched the total energy in the matter. For all higher temperatures, corresponding to earlier epochs and greater densities, the Universe was *dominated* by radiation, and matter played a secondary role. So we have our formula for reconstructing the conditions in the Big Bang – we know the present-day temperature of the Universe, the number of photons present for every proton or neutron, and the laws which tell us how conditions change as we wind the clock back. The simplest version of those laws, the simple Friedman–Lemaître cosmology, plus the known facts about the background radiation today, turn out to be precisely the recipe needed to cook 25 per cent of the original matter into helium in the first few minutes of the evolution of the Universe.

The cosmic fireball

How far back – how close to $t = 0$ – can we push the laws of physics to provide a working description of the Universe? The greatest density of matter occurring naturally in the world today is in the nucleus of an atom, where protons and neutrons are packed alongside each other, cheek by jowl. Nuclear reactions – reactions involving protons and neutrons – are responsible for the existence of the variety of chemical elements we see about us, and similar reactions shortly after the birth of the Universe established the proportions of hydrogen and helium that went as fuel into the first generation of stars. The standard model of the Big Bang derives from the work by Wagoner, Fowler and Hoyle which calculated how much helium could have been produced in the Big Bang, and it effectively tells the story of the evolution of the Universe from the time when the density of matter was about the same as the density of matter in an atomic nucleus today, or perhaps a little lower. The temperature at which this occurred was about 10^{12} K (1,000 billion K), the density was the density of nuclear matter (10^{14} grams per cubic centimetre), and the time was 0.0001 (10^{-4}) of a second after $t = 0$.

These conditions are so extreme that before we can look in detail at how the Universe developed from that state – from the Big Bang itself – we need to remind ourselves of the relevant laws of physics that describe such extreme conditions. One crucial point is that radiation plays a far more important role in the Big Bang than it does in the Universe today, and the reason is easy to see. If you picture the model of a universe being wound back in time and contracting, nothing very much happens to the individual atoms, let alone their nuclei, for a very long time. Because galaxies are so far apart from one another, it takes billions of years of contraction to bring them into contact. And even then the contraction still has a long way to go before individual stars are squeezed together into one amorphous lump. But the background radiation, although only a weak hiss with a temperature of 3 K, fills the Universe entirely today, and it always has filled the Universe entirely. As soon as the imaginary contraction begins, the radiation is affected, and its temperature begins to rise. By the time stars are at last squeezed closely enough together for individual atomic nuclei to begin to feel the effects, the density of radiation at every point in space has increased to the point where it carries far more energy than the energy stored up in particles. It is no longer 'background' radiation, but very much at the forefront of physical processes going on in the hot, dense universe.

The energy equivalent of a particle of mass m is, of course, mc^2. The relation $E = mc^2$ tells us that a sufficiently energetic packet of radiation (a photon) can convert into matter with the appropriate mass, and vice versa (there are also other rules which have to be followed, but this is the key constraint). At the high energies and densities of the Big Bang, it does indeed make sense to think of radiation in terms of particles; in fact, as is well known, in the strange world of quantum physics all particles can also be thought of as waves, and all waves can be thought of as particles. Energy and mass are equivalent and interchangeable, and so are the concepts of particle and wave. But a photon, a packet of energy, cannot just disappear and be replaced by a single particle. Particles come in pairs, each with a counterpart, called an antiparticle, that is in a sense a 'mirror image' of the particle. The mirror image of an electron is a particle called a positron, which carries a positive charge instead of the electron's negative charge, hence its name. If an electron meets a positron, the pair annihilate in a burst of high-energy radiation, gamma radiation. And a sufficiently energetic burst of gamma radiation can produce a *pair* of particles, a positron and an electron.

At the time we are talking about now, in the Big Bang between 10^{-4} and 0.1 (that is, 10^{-1}) of a second after $t = 0$, the Universe was dominated by radiation. You can think of this dominance in two ways. First, the density of the radiation (the amount of energy it contained in each small volume) was so great that there was an energy equivalent to a positron-electron pair (roughly speaking) in each volume of space corresponding to the size of a positron-electron pair. So the energy could happily switch from electromagnetic energy into electrons and positrons and back again. Or think of it in terms of particles of electromagnetic energy, photons. For every nuclear particle (every proton and every neutron) there were a billion photons, and each of those photons could, and did, change into an electron-positron pair, while positrons and electrons, meeting up in this primeval maelstrom, would annihilate and produce more gamma photons to replace the ones that were turning into electron-positron pairs. The fireball was dominated by photons, electrons and positrons, and by massless particles called neutrinos. Perhaps, however, 'fireball' is not the best term for the Universe at that time. At 0.01 second after $t = 0$, the energy density of the Universe, in terms of $E = mc^2$, was equivalent to nearly four billion times the density of water here on Earth. Some fireball!

The protons and neutrons (collectively dubbed nucleons, since they are the nuclear particles) were relatively stable even under those extreme conditions. Left on its own, a neutron will decay spontaneously in a few minutes, turning into a proton, an electron and a neutrino. But the timescale of the fireball involves fractions of a second, so a particle that is stable for several minutes is effectively eternal. The proton and the neutron have similar mass to one another, and this is a bit less than two thousand times the mass of the electron. So to make proton-antiproton pairs, or neutron-antineutron pairs, you need correspondingly greater energy density of radiation (more energetic photons). The required energy was available even earlier in the life of the Universe, before $t = 10^{-4}$ seconds, but the standard Big Bang, as developed in the late 1960s, only deals with events after the density of the Universe fell below the density of nuclear matter, and protons and neutrons condensed out of the radiation.

The final point that needs to be emphasized before we look at what happened as the Universe cooled still further concerns the timescale on which all these changes were happening. Today, the Universe as a whole doesn't change noticeably in 0.001 second, or even in 10 million years. Cosmologists say that the age of the Universe is between 10 and 20 billion years, and they are supremely untroubled about the range of possible values, which covers a factor of 2 (2×). But conditions changed more rapidly when the Universe was young, and fractions of a second become important in interpreting events in the Big Bang. A characteristic timescale, at any stage of the evolution of the Universe, can be thought of as the time it takes for any chosen region of the Universe to double in size (today, that would be equivalent to the time it takes for the distance between any two galaxy clusters to double). Gravity is continually slowing down the expansion of the Universe, so this significant time scale is itself increasing as the eons pass. It takes longer and longer for anything significant to happen to the appearance of the Universe at large. The corollary to that is that the closer you get to $t = 0$, the less time it took for significant changes to occur. The characteristic time is roughly proportional to the reciprocal of the square root of the density of the Universe at any epoch (the bigger the density the shorter the characteristic time), and at the start of the era dominated by photons, electrons and positrons, and neutrinos this time scale was a mere 0.02 second.

You can get a rough feel for how the important timescale changed as the Universe aged by thinking in terms of powers of ten, and working backwards in time. The age of the Universe is about 15 billion years, or in round powers of 10, 10^{10} years. Astronomers are happy with their estimates for the age of the Universe, because those estimates all agree to the same power of ten – there are no estimates as small as 10^9 years, and none as large as 10^{11} years. If we look back into the cosmic past, the first significant landmark might be at about 10^9 years, when the Universe was one-tenth as old as it is today, and would have looked noticeably different. The next landmark would be when it was a tenth younger still, at 10^8 years (100 million years), 1 per cent of its present age, and so on. In those terms, everything that happened in the interval from the first tenth of a second (0.1 second) to the end of the first second is about as interesting, and significant, as everything that happened in the interval from the first hundredth of a second (0.01 second) to the end of the first 0.1 second, and so on. The analogy is not precise, but it gives a flavour for the importance of the fast-changing world of the early Universe. And there is still another way to put this in perspective. The age of the Universe, in seconds, is a few times 10^{17} seconds. One second is 10^0. The interval from the present to the first second covers a span of seventeen powers of 10. If we travel back in time the same distance the other side of one second, we arrive at 10^{-17} second. In a sense, a very real sense, the interval from 10^{-17} second to 1 second is equivalent to the interval from 1 second to the present; and physicists now talk in terms of events that occurred back to within 10^{-40} second of $t = 0$, which in the same terminology lies 2½ times further back towards the moment of creation than we are from the time $t = 1$ second. In those terms, the events from 10^{-4} second up to about 4 minutes seem almost mundane – but those events shaped our Universe.

The first four minutes

The best-known description of the cosmic fireball from the era dominated by radiation, electron–positron pairs, and neutrinos onwards is Steven Weinberg's book *The First Three Minutes*. As Weinberg acknowledges in that book, the title is really a bit of author's licence.

His account of the Big Bang actually starts at $t = 10^{-2}$ second, 0.01 second after the moment of creation, and the main action he describes occupies the next 3 minutes and 46 seconds of the life of the Universe. He was writing in 1976, and at that time physicists had only the haziest of ideas about what happened during the first 0.01 second, so his starting point is reasonable enough. And the book still provides a very clear guide to what happened during those crucial 3¾ minutes.[2] So I shall follow Weinberg's now classic summary of how conditions changed during the 4 minutes in which the Universe was transformed from a uniform, very dense soup of radiation and matter into a mixture of about 75 per cent hydrogen and 25 per cent helium, with the radiation decoupled from the matter and left to fade away into the weak background that we know today.

The story begins with the Universe at a temperature of 100 billion K (10^{11} K) at time $t = 10^{-2}$ second. It is dominated by radiation, by the electron-positron pairs that are both produced by the radiation and which annihilate to produce radiation, and by the massless neutrinos and their antineutrino counterparts. The protons and neutrons that are so important to the material world today, and make up all the stars and planets, the clouds of gas and dust in space and the atoms in our own bodies, are at this time simply an insignificant component of the soup, their numbers a mere one billionth of the total number of photons. The nucleons are being constantly bombarded by the electrons, positrons and neutrinos, and this bombardment causes them to continually change their spots. An antineutrino colliding with a proton produces a positron and a neutron, while a neutrino colliding with a neutron produces an electron and a proton, and both of these reactions can run in either direction. *Individual* nucleons are constantly bombarded, and change repeatedly from neutron to proton and back again. But on average, as long as the energy of the fireball is great enough for all these reactions to proceed easily, there will be just about the same number of protons as there are neutrons in any specified volume of the Universe. Things begin to change, however, when the temperature drops to about 30 billion K.

Particle physicists often measure energy and mass (which is the same thing, if you allow for the factor c^2) in units of electron volts. One eV is the energy an electron would gain when accelerated across a potential difference of one volt. This is a pretty small unit. A typical photon of visible light carries an energy of about 2.5 eV, and the mass of an electron is 510,000 eV, just over 0.5 MeV.[3] The mass of a proton is 935 MeV, and the mass of a neutron is almost, but not quite, the same as the mass of a proton. That 'not quite' is the key to the next stage in the evolution of the Universe.

When the temperature of the Universe was as high as 10^{11} K, the typical energy carried by each electron, photon or other particle was about 10 MeV, 10 million electron volts. Some had more energy, some less, but this was a good average value. This is a lot less than the masses of the nucleons, which is why the nucleons were able to retain their identity at that time. And it is a lot more than the mass of an electron-positron pair, which is why such pairs could be created so easily at that time. But it is also a lot more than the *difference* in mass between a proton and a neutron, which is just under 1.3 MeV. To an electron or a neutrino carrying 10 MeV of energy, it made very little difference whether it reacted with a proton or a neutron, and the two key reactions for converting protons into neutrons and neutrons into protons went equally happily in each direction. But as the temperature of the Universe fell, the energy carried by each particle declined in proportion. With less energy available to drive the reactions, the mass difference between protons and neutrons began to be important, and it became relatively more difficult to trigger the reactions that converted the lighter protons into the heavier neutrons. The 'uphill' reaction could still occur, if a sufficiently energetic electron collided with a proton, but sufficiently energetic electrons became scarcer and scarcer, and were, from now on, significantly less abundant than the particles with slightly lower energy needed to convert neutrons into protons.

Just over one tenth of a second after $t = 0$, the temperature of the Universe was 3×10^{10} K, the energy density had fallen to 30 million times the energy density of water, the expansion rate had slowed so much that the characteristic time scale of the Universe was now 0.2 sec, and although the proportion of nucleons to photons was still a modest 1 in 1 billion, the proportion of neutrons to protons was no longer 50:50, but 38 per cent neutrons to 62 per cent protons.

About a third of a second after $t = 0$, a major change occurs in the

Universe. At the high temperatures of the early fireball, the particles were happily involved in many interactions, including an interchange between electrons, positrons and neutrinos by which an electron-positron pair could annihilate to produce a neutrino-antineutrino pair, and vice versa, as well as the nucleon reactions already mentioned. But neutrinos are very reluctant to interact with other matter under any conditions that we would regard as normal. They pass right through the Earth without being affected – indeed, neutrinos produced in nuclear reactions at the heart of the Sun stream out through the Sun itself without being significantly affected on the way. To neutrinos, ordinary matter is transparent. And 'ordinary matter' to neutrinos means anything less extreme than the conditions that existed a third of a second after the moment of creation. Then, or soon after, the neutrinos ceased to interact with electrons, positrons or anything else, but remained as a background sea (rather like the cosmic microwave background radiation, but far less easy to detect) filling the Universe but playing only a minor part in its evolution.[4]

So by the time the temperature had cooled to 10^{10} K (10 billion K), at $t = 1.1$ seconds the density was down to a mere 380,000 times the density of water, neutrinos had ceased to interact with matter (they had decoupled), and the characteristic expansion time of the Universe had stretched to 2 seconds, while the balance between protons and neutrons had shifted still further, with 24 per cent neutrons and 76 per cent protons. With the temperature continuing to fall, below 10^{10} K, photons carrying enough energy to create electron-positron pairs became increasingly rare, and during this phase of the evolution of the Universe electrons and positrons were annihilating one another faster than new pairs were being created.

From now on, the breathless pace of evolution is slowed to something almost familiar; we deal in whole seconds, not fractions of a second, and the particles and their reactions are very similar to the particles and reactions that provide the energy of the Sun and stars today.

By the time the temperature has dropped to 3 billion K (3×10^9 K), 13.8 seconds after $t = 0$, no more electron-positron pairs are being produced, and the ones that remain are being annihilated. Nuclei of deuterium (one proton plus one neutron) can form temporarily, but are knocked apart by collisions with other particles almost as soon as they do form. And although neutrons are still being converted into protons, with less energy available this reaction is slowing down dramatically, and still 17 per cent of the nucleons are in the form of neutrons. Three minutes and 2 seconds on from the moment of creation, the temperature of the Universe has cooled to 10^9 K, and at last we can compare this with something in the present-day Universe. The temperature at the heart of the Sun is about 15 million K; 3 minutes after $t = 0$, the Universe had cooled to a temperature only seventy times greater than this. The particle reactions that were so important a few minutes earlier have virtually ceased, but now the Universe is old enough for the natural decay of the neutron to become important, and in every 100 seconds from now on 10 per cent of the remaining free neutrons will decay into protons; the proportion of neutrons is already down to about 14 per cent. But they are saved from extinction as the temperature falls still further, to the point where deuterium nuclei can hold together.

Now, at last, the reactions described in outline by Gamow and his colleagues, and in detail by Wagoner, Fowler and Hoyle, can take place. Nucleosynthesis rapidly builds up nuclei of helium-4 but essentially stops there because, as we have seen, there are no stable nuclei with masses 5 or 8, and nucleosynthesis can bridge those gaps, as Hoyle explained back in the early 1950s, only under the conditions found inside stars – which don't yet exist.[5]

Once helium production begins, all the available neutrons are quickly bound up in this way, and they are then stable. This happens when the proportion of neutrons is about 13 or 14 per cent of the total number of nucleons, and in nuclei of helium-4 each neutron is accompanied by one proton. So the proportion of the total mass of nucleons converted into helium-4 is simply twice the abundance of neutrons when the reactions begin, about 26 to 28 per cent. Nucleosynthesis begins at a temperature of 900 million K (9×10^8 K) at 3 minutes and 46 seconds after the moment of creation. By $t = 4$ minutes, the standard model of the Big Bang has created the conditions that produce just the amount of helium observed in the Universe.

This great triumph of the standard model depends crucially on the fact that the reactions converting protons into neutrons and neutrons into protons 'froze' just when they did, so that a residue of 14 or 15 per cent neutrons was left at the time nucleosynthesis began. These critical reactions, and the point at which they freeze, are very sensitive not just to temperature but also to the rate at which the temperature of the early Universe was falling. If the freeze happened at an 'age of the Universe' of a few seconds, then the proportion of helium in the Universe is indeed just under 30 per cent. But if everything happened a little bit faster, and the freeze happened at 0.1 second, the proportion of helium produced by the Big Bang would be almost 100 per cent (because nucleosynthesis also gets going that much quicker), while if the freeze happened at 100 seconds, in a universe evolving that much more slowly, there would be no helium produced in the Big Bang, because all the neutrons would have turned into protons before nucleosynthesis could begin.

The rate of fall in temperature is specified by the standard model in its simplest form, and is tied in with the temperature of the cosmic microwave background today, which gives the crucially important estimate of 10^9 photons for every nucleon in the Universe. The ratio holds today, observations of the cosmic microwave background tell us, even though the photons are spread out through the Universe while the nucleons are concentrated in material lumps. So it must have held in the fireball stage of the Universe, when the radiation dominated the matter and drove the reactions in just the right way to produce the amount of helium we see. And the standard model also sets some constraints on the possibility of other particles existing in the Universe today. Neutrinos and antineutrinos interfere with the processes that convert protons into neutrons and neutrons into protons, so the success of the standard model also tells us something about the number and kind of neutrinos present in the Big Bang, and therefore left over for us to find now.

So we have a second successful prediction – or requirement – of the standard model to add to the astonishing discovery, in the 1920s, that the Universe is expanding. The requirement that there is three times as much hydrogen as there is helium in the Universe is a striking vindication of the simplest cosmological models of the Big Bang. The third leg of the tripod on which the standard model rests is the presence of the cosmic microwave background radiation; but to see exactly where that comes from, we have to move on from the moment of creation not in steps of a few seconds or even minutes, but in thousands and then billions of years.

The next 10 billion years

A little more than half an hour after $t = 0$ (at $t = 34$ minutes and 40 seconds, to be precise), almost all of the electrons and positrons have been annihilated, and the Universe has begun to resemble the empty state we find it in today. Almost, but not quite, all of the matter has gone. In addition to the billionth part of the number of photons that is present as nucleons, when the electron-positron pairs finally annihilate, just one electron in a billion is also left over, exactly the amount required to balance the positive charge on all the protons in the Universe, and to ensure that eventually the matter will settle out as stable, uncharged atoms, with every proton in every atomic nucleus matched by an electron in the cloud on the outside of the atom. Where does this tiny proportion of matter come from? Why isn't there a perfect symmetry between particles and antiparticles, so that everything annihilates and only radiation is left as the Universe cools? The answers emerge from an understanding of the world of particle physics under conditions even more extreme than those during the epoch of the life of the Universe following the first hundredth of a second, and they are among the simplest, but also most profound, of the puzzles resolved by the discoveries described in later chapters. For now, though, let's stick with the expanding, cooling fireball half an hour after the moment of creation.

By now, the temperature of the fireball is down to 300 million K, and the energy density of the Universe is only 10 per cent of the mass density of water. About 69 per cent of this energy is carried by photons and 31 per cent by neutrinos, and the expansion timescale appropriate at this time has stretched to 75 minutes. Although all the available neutrons have been cooked into helium nuclei, the Universe is still too hot for stable atoms to form – as soon as a positively charged proton or helium-4 nucleus latches on to a negatively charged electron, the electron is knocked out of its grip by an energetic photon. This is the 'radiation era' of the Universe, with no significant particle interactions to worry about, and with the remaining matter dominated by the radiation. It lasts for about 700,000 years, until the temperature drops to about 4,000 K and the nuclei and electrons are at last able to hold together against the ever-decreasing battering they receive from photons.

The time at which this occurs is not well defined. As early as 300,000 years after the moment of creation some hydrogen atoms are beginning to form and survive for a reasonable length of time without being ionized by the radiation; after $t = 10^6$ years, all of the electrons have been bound up in atoms, so efficiently that only one electron and one proton are left out on their own for every 100,000 stable atoms, and the 'decoupling' of matter from radiation is complete. From now on, radiation and matter scarcely interact at all, since although electromagnetic radiation and charged particles interact strongly, neutral particles, such as atoms, have little effect on radiation, or radiation on them. Like the sea of neutrinos that decoupled earlier on, the photons are left to fade away into a cosmic background.

The decoupling era, a little less than a million years after the Big Bang, marks the last time matter and radiation were closely involved with one another. So the cosmic background we see today is in effect a view of the Universe at that time. The fact that the cosmic background is uniform, isotropic and homogeneous tells us that the Universe as a whole was uniform, homogeneous and isotropic 700,000 years after $t = 0$. This is the closest direct observation of the Big Bang we have. But remember the primordial neutrinos. In principle, they might be detected, and the equations of the standard model tell us that they should form a background sea filling the Universe today with a temperature 70 per cent of the temperature of the photon background, about 2 K. And they decoupled only just over a second after $t = 0$. If these neutrinos are ever detected, they would provide the most dramatic confirmation yet of the accuracy of the standard model, and would give us a view, distant though it is, of the Universe when it was one second old.

Just before the decoupling of matter and radiation, the entire Universe resembled the surface of the Sun. It was hot, opaque and filled with a yellow light. As matter and radiation decoupled, it suddenly became transparent, and at about the same time the energy density of the radiation fell below the equivalent density of the matter in the Universe. From about $t = 10^6$ years onwards, the Universe has been dominated by matter and by gravity. And in round terms you can get some sort of feel for how long ago that was in terms of the redshift. The greatest redshifts ever measured for any astronomical objects are those of a few quasars, with redshifts (z) of a little over 4.[6] The redshift of the decoupling epoch and of the epoch when matter began to dominate the Universe each correspond to about $z = 1,000$. The wavelength of each photon in the cosmic microwave background has been stretched by a factor of 1,000 since it last interacted with matter.

Although the radiation era seems at first sight to have been one in which nothing much happened, compared with the first four minutes, it was probably at this time that the irregularities which later grew up to become galaxies and clusters of galaxies first developed. At the end of the radiation era, when stable atoms had just formed, there were about ten million atoms in every litre of the Universe. Today, there is only one atom, on average, in every thousand or so litres of space. The number density of atoms at decoupling was at least a thousand times greater than the density of even a galaxy today, so, clearly, galaxies as we know them must have formed after decoupling. But the matter that became the dominant feature of the Universe probably inherited irregularities from the radiation era. In some places, the density was already slightly greater than in other places. By the time matter was dominating the evolution of a transparent, dark and cooling Universe, it was already grouped into clumps that, because of the insistent pull of their own gravity, did not thin out as rapidly as the Universe at large. Within such clumps of matter with above-average density, some regions formed clouds of gas that began to break up and collapse, eventually forming the stars of our own Milky Way and other galaxies. By the time half the present age of the Universe had elapsed, our own Milky Way Galaxy was in existence in much the form we see it today; 4.5 billion years ago, our Sun and its system of planets were in existence, having formed out of interstellar material that had already been processed and reprocessed in the interiors of many stars, and contained an enrichment of heavy elements as well as its inheritance of hydrogen and helium from the Big Bang.

For most of the past 10 billion years, most of the nucleonic matter in the Universe has been bound up in stars and galaxies, with the only large-scale change being the steady separation of clusters of galaxies from one another as the Universe expands, and the steady cooling of the ever more redshifted background radiation. But the details of galaxy formation remain obscure, and there are rival theories, put forward by different groups of astronomers, to explain how the matter in the Universe got to be grouped into the patterns we see it in today. Once again, it turns out that the latest understanding of particle physics provides a clue. Indeed, the answers to most of the questions left unanswered by the standard model turn out to lie in a better understanding not of the way the Universe is today, but of the way it was in the first millisecond, *before* the Big Bang as described by the standard model.

Remaining questions

In spite of the enormous success of the standard model in the 1960s and 1970s, it left a handful of unanswered questions. Where did the tiny proportion of matter (compared with the number of photons) come from? Why is the Universe so extraordinarily uniform and homogeneous, and why is the density of the Universe so close to making it flat? What happened in the interval from $t = 0$ to the end of the first millisecond? How did the Universe come into being – what *did* happen at the moment of creation itself? And, at the other end of the time scale of cosmic evolution, what will the ultimate fate of the Universe be?

These remaining big questions of cosmology, concerning the very origin of the Universe and its fate – the beginning and end of time – may have been answered by a single highly successful theory that emerged during the 1980s. This theory, which goes by the name of inflation, also explains the incredible flatness of the Universe.

At last we have come down to the metaphysical nitty-gritty, the questions that used to be regarded as beyond the scope of science. But now, science knows no limitations, and can tackle all of these questions, even if the answers are not yet complete, or completely understood. We even have a line of attack on the problem of the creation itself. Understanding of all of these deep mysteries depends on getting a handle on what happened in the tiny fraction of a second before the time where the standard model picks up the story – before, in that sense, the Big Bang itself. As cosmologist Ted Harrison, of the University of Massachusetts, has commented, 'more of cosmic history occurs in the first thousandth of a second than has occurred in 10 billion years since' (*Cosmology*, page 354). But the key insight into the true nature of the Universe actually comes from a re-examination, with the aid of the latest observational techniques, of an old puzzle – the question, going back to the time of Hubble himself, of just how fast the Universe is expanding today, and what this can tell us about the matter contained in the Universe. It is time to look more closely at what we know about the key number in cosmology, Hubble's constant itself.

Chapter 8

Close to Critical

The most important observation in cosmology is that the light from all galaxies outside the Local Group is redshifted, which implies that the Universe is expanding and has evolved from a much denser state, the Big Bang. Hubble showed that for as far as it is possible to estimate distances to galaxies by other means (studies of variable stars or bright clusters of stars in galaxies, and so on) the redshift is proportional to distance. There seems no reason to doubt that this rule applies to all distant galaxies, so distances to galaxies far beyond the range of the variable star technique, or any other trick, can be determined simply by measuring the redshift in their light and multiplying by the constant known today as Hubble's constant, H. Indeed, even if we do not know the exact value of H, as long as we know the rule 'distance equals constant × redshift' we can determine the *relative* distances to galaxies – that one is twice as far away as another, while a third is, say, 12 times further away than the first. And that is just as well, because estimates of the exact value of H have changed significantly since Hubble's day, and even now there are two schools of thought regarding the value that ought to be assigned to Hubble's constant. Because all our estimates of extragalactic distances are linked to Hubble's constant, the effect is that one school holds that the observed Universe is twice as big as the estimate favoured by the other school. Yet both groups base their estimates on the same data, and each rejects the other claim as impossible. These conflicting claims first emerged at a major scientific meeting in Paris in 1976, and are still unresolved. But as we shall see, only one of them is fully compatible with the possibility that our Universe is closed, curved around on itself to form a finite but unbounded whole. And there is an increasing weight of independent evidence that the Universe is indeed self-contained in this way.

Stepping stones to the Universe

In comparison with the precision with which physicists today know such fundamental numbers as the mass of the proton, or the size of the constant of gravity, it may seem remarkable that there could be as much uncertainty as a factor of two in our knowledge of the distance scale of the Universe. But this apparent vagueness becomes a little more understandable when you recall that, as we saw in Chapter 3, it was only in the 1920s that astronomers appreciated for the first time that there is more to the Universe than our own Milky Way Galaxy, and began to estimate distances to other galaxies. Indeed, compared with the best estimates available in 1929, the Universe today is ten times 'bigger' than it was – at least, it's ten times bigger than astronomers *thought* it was.

The main reason for the imprecision of their estimates is simply that you can't put the Universe in your laboratory to study it. A proton can actually be manipulated in the lab and its properties measured; but our knowledge of the Universe depends on observations of faint and distant objects, and is always at best secondhand. The wonder is that any plausible numbers for such properties as the distance to galaxies and quasars emerge at all, and the ultimate parameter, the distance scale that gives us the size of the Universe, is reached only by using a series of scientific stepping stones, each of which can only be reached with the aid of earlier steps. A mistake anywhere in the chain of reasoning throws off all the calculations down the subsequent steps.

The shorthand expression 'size of the Universe' is something of a misnomer. What astronomers are interested in is the bit they can see, with the aid of telescopes and other instruments, and what they want to know is a way to calculate the distance to every galaxy and other object they see out beyond our own Milky Way. They prefer to talk about the distance *scale* of the Universe, the relative distances between galaxies, precisely because these relative distances stay the same whatever the actual value of H.

Hubble's constant is the key number in all of cosmology. Armed with an accurate value of H and redshift measurements, it would be possible to calculate the distance to any galaxy. And it is the precise value of H that has been bitterly disputed by the experts for ten years. Allan Sandage, of the Mount Wilson and Las Campanas Observatory, and his colleague Gustav Tamman of the University of Basel, estimate it as 50 kilometres per second per Megaparsec (km/sec/Mpc). Gerard de Vaucouleurs of the University of Texas advocates 100 km/sec/Mpc). Neither seems willing to budge. But even within that range of possibilities H is telling us a great deal about the Universe we live in.

The time that has elapsed since the Big Bang depends on how fast the Universe is expanding – on Hubble's constant. So measuring Hubble's constant also gives us, immediately, an estimate of the age of the Universe. That estimate is always too big, as Figure 5.5 shows, because gravity must have slowed down the expansion as the Universe has aged, making H smaller today than it was in the past (which is why it is sometimes denoted by H_o, denoting the value of Hubble's 'constant' today, and why some pedants prefer the term 'Hubble parameter', since it is not really a constant). The rate at which the universal expansion is slowing down depends, of course, on how much matter there is in it. The more matter there is, the more strongly gravity is acting to halt the expansion.

The density of matter in the Universe is denoted in cosmology by Ω, the Greek capital letter omega. This parameter is defined in such a way that if the cosmological omega is less than one, the Universe is open and will expand forever, while if it is bigger than one the Universe is closed and must inevitably end in a Big Crunch (sometimes called the 'omega point') like the Big Bang in reverse. If the density of the Universe is just the minimum required for closure, and omega has the critical value of one, then the true age of the Universe, the time that has elapsed since the Big Bang, is exactly two-thirds of $1/H$.

Even if we do not know the precise value of H at least the limited range of possibilities is telling us something about the time that has elapsed since the Big Bang. The inverse of H is called the Hubble time, and, once all the kilometres and Megaparsecs have been divided into one another, and the seconds converted into years, this ranges from 10 billion years (for $H = 100$ km/sec/Mpc) to 20 billion years (for $H = 50$ km/sec/Mpc). The corresponding range of possible ages for the Universe if omega is equal to one is, in round terms, 6.5 billion years to 13 billion years. And the uncertainty arises because of the difficulty for astronomers, described in Chapter 3, in finding the accurate distance to just one galaxy outside the Local Group.

The stepping stones out into the Universe were, as we have seen, parallax, the moving cluster method which gave the distance to the Hyades cluster of stars, and the brightness-magnitude (Hertzprung–Russell, or HR diagram) technique which used the Hyades cluster as a whole to calibrate the brightnesses, and therefore the distances, of other clusters. The key step *outside* our own Galaxy came when Cepheid variables were found to provide 'standard candles' – stars whose intrinsic brightness could be inferred from the length of the cycle of their periodic fluctuations. But even Cepheids only take us out to the nearby galaxies, about 5 Mpc. Stellar explosions, novae and supernovae, can be used as distance indicators to more remote galaxies. A supernova shines, briefly, as brightly as a whole galaxy of ordinary stars, and the apparent brightness of such a flare tells us how far away the galaxy in which the exploding star sits is – provided, of course, that we have some idea of the absolute brightness of a typical supernova explosion. Even then, supernovae are far from common. So astronomers have to fall back on secondary distance indicators for any except the closest galaxies.

Secondary techniques are much less reliable. First, astronomers study the properties of those galaxies for which distances are known, and try to find common features. Then, by comparing those features with the equivalents in more remote galaxies, they estimate the distances to those galaxies.

For example, many spiral galaxies contain large clouds of ionized hydrogen, called HII regions. If all HII regions are the same size, and if it is possible to measure the diameters of these clouds using radio astronomy techniques, then the distance to a galaxy that contains such clouds can be found by comparing their apparent sizes with those of clouds in a nearby galaxy. There are plenty of 'ifs' in that chain of reasoning, and other secondary methods are no better. Which is why it has proved possible for Sandage and de Vaucouleurs to hold such differing views on the exact value of Hubble's constant.

There are several reasons why the two schools of thought disagree.

First, de Vaucouleurs assumes that when we look out through the polar regions of our own Galaxy at distant galaxies then the light we see from them is dimmed slightly due to obscuring dust. Sandage disagrees, so his estimates of galaxy brightnesses, and distances, differ from de Vaucouleurs'. Second, Sandage has moved away from the simple period/luminosity relation for Cepheids found by Leavitt, and recognizes slightly different period/luminosity relations for Cepheids of different colours – an effect which de Vaucouleurs ignores. Among other differences, Sandage uses an estimate of 40 pc for the distance to the Hyades, much lower than the figure used by anyone else – and it is the distance to the Hyades which, through the main sequence method, gives the distance to the first Cepheid used to calibrate all the others. The two experts already disagree by more than 30 per cent in their distance estimates in our own astronomical back yard, and the discrepancies get worse as they move out beyond the nearer galaxies. Part of the further discrepancy in the estimates for distances out across the Universe lies in different interpretations of how much the cosmological motion of our Local Group is distorted by the pull of the Virgo Cluster. Redshifts, and distances, to other galaxies have to be corrected for this local effect before we can get a true picture of how rapidly the Universe at large is expanding, and the two groups disagree on the size of the correction required.

There are new techniques being developed which may resolve the controversy. When a supernova explodes, it blasts out a shell of material, expanding very rapidly. The light from the supernova actually comes from this expanding shell, and the Doppler shift in that light tells astronomers how fast the shell is moving, which makes it a straightforward matter to calculate how big the shell is a certain time after the initial outburst. If there were some way to measure the apparent size of such a shell then this could be related to calculations of its actual size to give a direct, and theoretically well founded, primary indication of distance.

The idea is simple, but the practical application is tricky. At the distance to the Virgo Cluster, for example, we are talking about measuring an angular size smaller than one millionth of a degree. Nevertheless, this astonishing accuracy is now being achieved by radio astronomers using the technique of Very Long Baseline Interferometry

(VLBI). The first successful application of the technique in this way was announced in 1985, and gave the distance to a supernova in a galaxy called M100, 19 million parsecs away. The observations of the growing supernova shell suggest that the value of H is about 65 km/sec/Mpc, which probably would not please either Sandage or de Vaucouleurs, if taken at face value, but the uncertainties involved in this first application of a new technique are very large, and certainly encompass the range of values that would please either side in the debate. The method could, however, become one of the most reliable ways of estimating distances to galaxies, and thereby the distance scale of the Universe, in the next decade.

How else could the numbers be checked? The most promising line of attack for the immediate future is simply to carry out traditional observations of Cepheids in more remote galaxies. The Hubble space telescope now provides enough resolution to pick out individual Cepheids in the Virgo Cluster, but has not yet given a definitive answer. Sandage and de Vaucouleurs cannot both be right. They may both be wrong. But there are some quite separate, powerful arguments in favour of the smaller value of H, and larger age, for the Universe.

The age of the Galaxy

The first estimates of H gave an 'age of the Universe' which was less than the age of the Earth, deduced by geologists. This conflict provided a powerful incentive for astronomers to find out what was wrong with their estimate of the age of the Universe, since the Universe itself must, clearly, be older than any star or planet in the Universe. The present range of estimates for H gives a range of possible 'ages of the Universe' which are ample to accommodate the known lifetime of the Sun and Solar System, which is now thought to be about 4½ billion years. But there are much older stars and star systems in our Galaxy, and the oldest of these have been around for long enough to rule out straightforward versions of the cosmological models with large values of H and with as much matter as there now seems to be in the real Universe.

Astronomers believe that they have a good understanding of

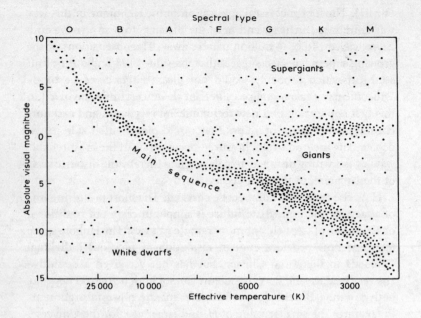

Figure 8.1 The H–R Diagram
The appearance of a star can be described in terms of its brightness (magnitude) and its temperature or colour (spectral type).
A Hertzprung–Russell diagram is a plot, like a graph, in which
the position of each star is determined by these two properties.
Most stars, burning their nuclear fuel in accordance with the
simple laws of physics, lie on a band called the main sequence.
Large, hot stars are to the top left of the main sequence; small,
dim stars lie to the bottom right.

how a star works. Their understanding of the processes of nuclear
fusion inside stars helps them to understand the nature of the HR
diagram, the relationship between the colour of a star and its brightness that is so useful in helping to determine distances inside our
Galaxy (Figure 8.1). The diagonal band of bright stars in the HR
diagram corresponds to stars like our Sun, which are young enough
to be 'burning' hydrogen into helium in their hearts. Stars with
different masses, but all busily burning hydrogen, sit along the main

sequence band of the HR diagram. When their hydrogen fuel is exhausted, however, their appearances change, in a way that can be thoroughly explained by computer models of how stars work, and which can be understood in outline from a few simple physical arguments.

In the heart of an ageing star, at the end of its life as a member of the main sequence, there lies a core of helium surrounded by a shell in which hydrogen is still being converted into helium. The shell spreads outwards, and the core gets bigger, as the star ages. The helium core itself contracts under its own weight and heats up, until eventually it becomes hot enough at the centre for a new phase of nuclear burning to begin, with helium now being converted into carbon. This will happen to our Sun in about five billion years from now. With a small, hot core pouring out even more energy than the Sun does today, the effect on the outer layers of the star will be to make it swell up, engulfing Mercury and Venus, and frying the Earth itself. The temperature at the surface of this huge ball of gas will be much less than that of the surface of the Sun today, so that the star will have a cool, red colour – such a star is known as a red giant, and many red giants are known to astronomers.[1]

These changes can be seen taking place in the stars of a cluster, when their brightnesses and colours are plotted in an HR diagram. The main sequence runs diagonally across the diagram, from top left to bottom right. Red giants lie above the main sequence, in the top right of the diagram. And although these changes take too long for an individual star to be observed changing its position across the diagram, the computer models tell us exactly how the red giants got there.

Stars with more mass burn their nuclear fuel more quickly, and shine more brightly, than stars with lower mass. They have to, simply to hold themselves up against the inward tug of gravity. Such high mass stars lie on the main sequence at the upper left of the HR diagram. As their hydrogen fuel is exhausted, they 'move' upward and to the right, off the main sequence. And as time passes all the stars in the main sequence peel away to the right, starting with those at the top left and finishing with those at the bottom right. When astronomers study clusters of stars in our galaxy, this is exactly what they see – a main sequence starting out happily enough from bottom

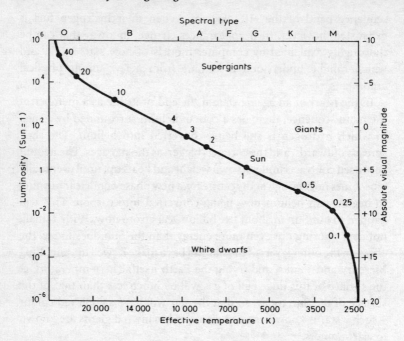

Figure 8.2 Concentrating on the main sequence, we can see how the position of our Sun compares with those of other stars. Luminosities are given in terms of the brightness of the Sun; numbers along the main sequence are the masses of the appropriate stars in terms of the mass of the Sun.

right, with low mass stars, but ending at some point and turning off to the right, into the red-giant branch. If we know the distance to a particular cluster, we can compare the stage of evolution it has reached against the standard computer models simply by measuring the point where this turnoff occurs, and that immediately gives an estimate for the age of the cluster.

As always in astronomy, there are uncertainties in the practical application of the technique. Corrections have to be made for the effects of interstellar dust on the light from stars as it crosses space to us; the turnoff point on the main sequence is never as precisely defined as I have made it sound; and there are other difficulties. With

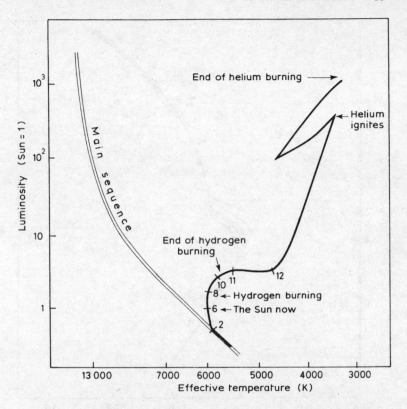

Figure 8.3 As a star like our Sun ages, it leaves the main sequence and becomes larger but cooler. Numbers along the solid line indicate the age of a one solar mass star in billions of years since its formation; our Sun has, in fact, already begun to leave the main sequence and is four or five billion years old.

all these uncertainties, however, it is clear that the ages of the oldest clusters of stars in our Galaxy lie in the range from 14 billion to 20 billion years, always assuming that the standard models of stellar evolution are indeed correct, with a best estimate of about 16 billion years.

There are other techniques which give similar ages for objects within our Galaxy. Radioactive isotopes found on Earth and in meteorites, for

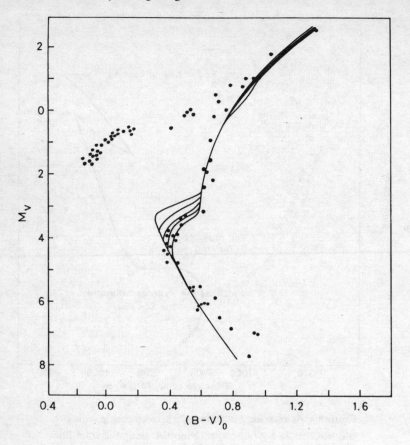

Figure 8.4 When astronomers study a group of stars that formed together and are all about the same age (a cluster), they can obtain a good idea of how old the stars are by measuring the point at which stars are leaving the main sequence. Massive stars, which are brighter and hotter than less massive ones, are the first to turn off in this way. In this case, the cluster is M92 and the five curves modelling the turnoff represent calculated ages of 10, 12, 14, 16, and 18 billion years. The age of the cluster is interpreted to be between 14 and 16 billion years. Only about ten clusters have well-determined ages, but these estimates are crucial to an understanding of the age of the Universe.

example, are thought to have been produced by supernova explosions within our Galaxy. These radioactive isotopes are unstable, and decay into stable elements in accordance with very precise rules that are well known from laboratory studies. The proportion of different radionuclides, as they are called, made in supernova explosions can be calculated using the same techniques which have proved so success-ful in explaining how stars work, and how they manufacture heavier elements out of hydrogen and helium. So the proportion of each kind of radioactive material left in the Solar System today can be used to determine the age of the Galaxy, assuming supernovae were manufac-turing radionuclides at a steady rate from the time the Galaxy formed until the Solar System was born. That gives an age of about 15 billion years, in fair agreement with the other evidence.

This is not quite enough to rule out de Vaucouleurs' model with H equal to 100. In that model, the maximum age of the universe is just 10 billion years, and that would correspond to an almost empty universe, with very little matter around to slow down the expansion. The difference between 10 billion and 15 billion is not enough to settle the argument, given all the uncertainties involved. But the discrepancy certainly looks uncomfortable if there is enough matter around to close the Universe, when, if $H = 100$, the implied age is only 6.5 billion years, less than half the age of the oldest known stars. Sandage's estimate, $H = 50$, implies an age of 13 billion years if the Universe is just closed, close enough to the estimates of stellar ages to make us feel more comfortable. At a meeting of the Royal Society in March 1982 where element creation was discussed in the context of the Big Bang, Roger Tayler, of the University of Sussex, summed up the age 'problem' by saying that 'if omega equals one, it is necessary that H_o is about 50 kilometres per second per Megaparsec,' and nothing has happened since 1982 to change that interpretation of the evidence. The agreement is better still if H is just 40 kilometres per second per Megaparsec, when the age of a closed Universe rises above 16 billion years – and such a value of H certainly cannot be ruled out, using Sandage's figures. The other way to get near-perfect agreement, of course, would be if the oldest star clusters in our Galaxy were actually a little younger than present estimates suggest. Could those estimates of stellar ages be just a *little* too high? Maybe, this time, the cosmology is telling the astrophysicists how best to refine their theories!

All these arguments would be resolved if cosmologists were able to measure the rate at which the expansion of the Universe is slowing down today. That would tell us, once and for all, just how much matter the Universe contains, whether it is open or closed, and which value of the Hubble parameter is closer to the truth. Unfortunately, although such measurements are possible in principle, it seems unlikely that they can be achieved in practice in the immediate future.

The redshift test

Hubble's law, the foundation stone of cosmology, is actually an imperfect description of the expanding Universe. That might seem like a flaw; but cosmologists are cunning enough to be able to turn the deviations from the straightforward Hubble law to their advantage, if only they could get enough good observations of very distant objects. Unfortunately, the observations, as yet, are not up to the task. This is because the law that velocity (redshift) is proportional to distance *is* so good in our immediate neighbourhood, and almost as far out as we can see across the Universe. It is where the law starts to need amending that things get interesting, but that only happens so far from home that we cannot yet be quite sure just which amendments are necessary.

The reason things get interesting has to do with the geometry of space. On the surface of the Earth, an architect designing the floor plan of a building can happily use the geometrical rules laid down long ago by Euclid, which strictly speaking apply only to a flat plane, and doesn't have to worry about the curvature of the Earth. We all learned Euclidean geometry in school, and remember that the angles of a triangle, for example, always add up to 180°. But if a team of surveyors were to lay out a series of very large, perfect triangles on the surface of a great, 'flat' desert (perhaps somewhere in the Sahara) and then carefully measured the angles of those triangles, they would find that the angles always added up to slightly more than 180°, and that the difference from 180° was bigger for bigger triangles. This is because the surface of the Earth is actually curved, forming a closed surface that is very nearly spherical, and the 'correct' geometry to apply in such cases is not the geometry of Euclid. Before, we were interested in how remarkably flat the Universe is; now, it is time to see what the tiny deviations from flatness can tell us about its probable fate.

If space itself is curved, then the deviations of its geometry from the everyday Euclidean geometry we know so well will also show up over suitably large distances. 'Suitably large', in this case, means over distances in excess of a few Megaparsecs – 10 million light years or more. This in itself is a significant feature of our Universe. It tells us that the geometry of spacetime is very nearly flat, and this in turn implies that the density of matter in the Universe is close to the critical amount required to make it closed. In principle, we can tell just how close by doing the equivalent of measuring the angles of huge triangles. In practice, however, we lack the skill to distinguish which side of the dividing line between the open and closed possibilities the Universe sits.

Measurements of the deviation of the Hubble law from the simple 'redshift equals constant × distance' are difficult. This, after all, is the law that is used to calculate the distances to remote galaxies, by measuring their redshifts! How else can we estimate the distances to far-away galaxies, in order to *compare* these with the redshifts, and see just how far out across the Universe the Hubble law actually works? If all galaxies were the same brightness, there would be no problem. The relative distances of all the galaxies could be worked out simply by ranking them in order of their brightness on the night sky – the dimmer they appeared, the further away we would know they must be. In fact, the situation would be a little more complicated than it seems at first sight, even if all galaxies were the same brightness. Where Euclidean geometry applies, the brightness of each galaxy falls off in proportion to one over the square of its distance – a galaxy twice as far away seems only one quarter as bright. This simple rule itself has to be modified when the geometry is different, and corrections ought, strictly speaking, to be calculated for each type of cosmological model. But that is the least of the problems cosmologists face when trying to apply this test.

It is only too clear from studies of galaxies that they are not all the same brightness. And it is doubtful whether the trick can be applied at all to the objects with the largest redshifts, the quasars. Quasars are thought to be the highly active, bright cores of galaxies. They shine even brighter than ordinary galaxies and are visible further away, at higher redshifts, where the geometric effects should be clearer. But there is no evidence that quasars all have the same brightness, so the trick cannot be worked. One team of astronomers, working at the Lick Observatory in California, has tried to use the brightness test on quasars which have similar spectra to one another, and might therefore be expected to have similar brightnesses. For what it is worth, this comparison suggests that the Universe is closed, and will one day recollapse. Nobody, however, is prepared to take this rather speculative interpretation of the quasar evidence at face value. Cosmologists are forced to restrict the technique to studies of galaxies, which are understood much better than quasars are, and where there is some hope of making the trick work.

Allan Sandage and his colleagues have carried out a long and patient study of galaxies and found some which do seem to have the same brightness as each other, and can be used as standard candles. Galaxies come in clusters, and very often the brightest galaxy in a cluster is a very large elliptical (so-called because of its shape, like a fat cigar; our own Milky Way Galaxy is a spiral, like the surface of a whirlpool, or the pattern made by cream stirred into a cup of coffee). As far as the observations can show, in the region of space out to a few Megaparsecs where the Hubble law holds precisely, the brightest large elliptical galaxy in any cluster is much the same brightness as the brightest large elliptical in any other cluster. There seems to be some natural maximum brightness such a galaxy can have, and any decent-sized cluster will have one galaxy that reaches the brightness limit. So, by using only these particular bright galaxies and plotting *apparent* brightness (in effect, distance) against redshift, Sandage is able to see how far the resulting plot deviates from a straight line, and thereby determine how rapidly the expansion of the Universe is slowing down.

There are still problems. Remember that when we look at light from a galaxy 10 million light years away we see the galaxy as it was 10 million years ago. Can we be sure that the brightness of the galaxy has not changed during that time, as the galaxy and the Universe have evolved? Over this sort of range, there probably has not been much change. But for more distant galaxies, which we see in their youth, there is every likelihood of change. Astronomers would like to allow for this in their calculations, but have no independent means of telling how much brighter (or, conceivably, dimmer) galaxies were when the Universe was young. Some of the experts try to make educated guesses and allow for this luminosity evolution; others prefer to leave the observations alone, since any correction they try to make might very well be in the wrong direction.

Having picked their way through this observational minefield, cosmologists then have to compare the observations with their theoretical models. They calculate the deviations from the simple Hubble law in terms of a deceleration parameter, often labelled q, which is defined in such a way that $q = \frac{1}{2}$ corresponds to $\Omega = 1$. At one time, Sandage's plots of redshift against brightness seemed to favour a value of q of about 1, implying that there might be twice as much matter in the Universe as the minimum needed for closure; his latest plots, however, with more data gathered over the years, suggest that this early assessment was over-optimistic. Today, the best that this technique can do is to tell us that q probably lies somewhere in the range from 0 to 2, and that on these grounds alone the open Universe models cannot yet be ruled out.

A more recent survey of galaxy redshifts was carried out in the late 1980s by Edwin Loh and Earl Spillar, of Princeton University. They reported a study of 1,000 galaxies that suggested the density parameter Ω is indistinguishably close to one. It is too early yet for this work to be regarded as definitive, but simply because it does depend on 'old-fashioned' astronomy with optical telescopes it carries a powerful message to the observers that perhaps they should have been paying more heed to the theorists, who, as we shall see in Chapter 9, have been pressing the case for a just closed Universe.

The new test is one of several that can, in principle, be based on the geometry of the Universe. If galaxies (or clusters) are distributed uniformly through the Universe, and geometry is Euclidean, then the number of galaxies we ought to see at different redshifts (different distances) can be calculated by the rules of geometry we learned at school. Broadly speaking, equal volumes of space ought to contain

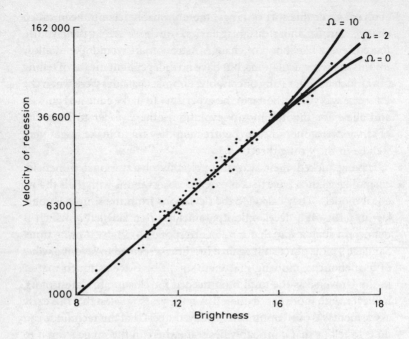

Figure 8.5 The Redshift Test Revisited
Because we see more distant galaxies as they were when the
Universe was younger, a comparison of their redshifts with those
of galaxies that are nearer to us, and therefore look brighter on
the sky, would in principle reveal how quickly the expansion of
the Universe is slowing down. Unfortunately, in practice these
observations can only indicate that the Universe is very nearly
flat. They can only tell us that the value of omega is roughly
between 0 and 2. (All the calculated curves overlap for small
values of redshift and brightness.)

equal numbers of galaxies. If the geometry is non-Euclidean, however,
when we try to determine equal volumes calculated in accordance
with Euclid's rules, and then count the numbers of galaxies in them,
there will be a discrepancy. 'Equal volumes' far from us will contain
either more or less galaxies than equivalent Euclidean 'equal volumes'
closer to us, and the exact number of galaxies more or less than the

number predicted by Euclidean geometry would tell us the fate of the Universe. In particular, if $\Omega = 1$ and the Universe is not expanding, the geometry is Euclidean; and if $\Omega = 1$ and the Universe *is* expanding, we will see a certain, precise deviation of the 'number counts' from the Euclidean prediction, because galaxies have moved apart as the Universe evolves.

Such deviations in the number counts are observed for many kinds of astronomical object – including radio galaxies and quasars – but it has proved very difficult to interpret the counts unambiguously. All of the studies of the large-scale dynamics of the Universe tell us that the geometry is very nearly Euclidean, that the expansion of the Universe is indeed slowing down, and that the amount of matter in the Universe must be fairly close to the amount needed for closure; but only the new study by Loh and Spillar tells us just how close to the dividing line we sit.

The Princeton team used the latest sensitive detectors to look at five tiny patches of the sky, each measuring about 7 arc minutes by 10 arc minutes (the Moon is 30 arc minutes across), and measured the redshifts of every galaxy they could detect in each patch. Because redshift is a distance indicator, they were, in effect, counting the number of galaxies in each of five pyramid-shaped volumes stretching out into the Universe from our Milky Way. By comparing the number of low redshift galaxies with the number of high redshift galaxies they could determine the geometry of each pyramid, without getting embroiled in the problems of the absolute brightness of each galaxy. Each of the five pyramids contained about 200 detectable galaxies, and the survey stretched out to distances of about a thousand Megaparsecs – or, put another way, they were looking back in time about one-fifth of the way to the Big Bang itself.

Just how you interpret these number counts depends on exactly what kind of mathematical cosmological model you choose. Loh and Spillar chose the simplest, the version of the equations of relativity developed by Einstein and Willem de Sitter in 1932. Loh and Spillar found that their counts of galaxies at different redshifts can best be explained by the simplest Einstein–de Sitter model, with zero cosmological constant and a value of omega of 0.9, with technical 'error bars' of ± 0.3 – indistinguishably close to one. It doesn't matter what the material holding the Universe together is, or even whether it is clustered in clumps the way galaxies are. It just has to be there.

I'll mention one other new technique for measuring the rate at which the Universe is expanding, since this looks a particularly promising line of attack for the future. It uses the curvature of space, but doesn't involve any dynamics. Astronomers have discovered a handful of objects where the light from a distant quasar is bent around a galaxy in the line of sight between us and the quasar, because of the way the gravity of the galaxy distorts spacetime in its vicinity. The effect – just like the light bending during an eclipse, but on a much bigger scale – produces a double or triple image of the quasar as seen from Earth, and because light takes longer to reach us round one side of the intervening galaxy than the other, when one image changes its brightness, or flickers, the other may not do so for years, and then flickers in exactly the same way as the equivalent image of the quasar reaches us by the other route. By comparing the way these images change, and measuring the time delay, cosmologists can calculate how long it takes for the light to reach us from the quasar, and thereby deduce the distance to it independently of any redshift measurement. The first application of this technique gave a value for H of 75 km/sec/Mpc – smack in between the values favoured by Sandage and by de Vaucouleurs, and bringing little pleasure to either camp. But by 1991, researchers from Brandeis University, in Massachusetts, and MIT had improved the technique by analysing a total of 11 years of observations of the first known 'double quasar', using the Very Large Array (VLA) radio telescope system. They found that the geometry of the lensing system tells us that if all the matter in the intervening galaxy that is bending the light is in the form of the bright stuff we can see with ordinary optical telescopes, then the Hubble constant has the value, in the usual units, of 46 if omega is zero and 42 if omega is one. In both cases, the 'error bars' are ±14. But if the intervening galaxy is embedded in a halo of dark stuff that we cannot see, the appropriate value of H_0 is 69 for omega equal to zero and 63 for omega equal to one, now with error bars of ±21.

This begins to point suggestively, if not yet completely persuasively, in favour of the value for H_0 favoured by Sandage and Tamman (implying a larger, older Universe), since there are, as we shall see in the coming chapters, compelling reasons to believe both that there is dark stuff surrounding all galaxies and that, in line with the observations made by Loh and Spillar, omega is very close indeed to one.

The work by Loh and Spillar, using the traditional technique of redshift measurements, caused a great deal of pleasure to one group of astronomers who had been tackling the same problem from a very different angle. They, too, had reached the conclusion that $\Omega = 1$, and had presented their conclusions a few months before the work of Loh and Spillar appeared. Those conclusions were initially regarded with deep suspicion by astronomers unused to the new tools of infrared astronomy from satellites. But this dynamic measurement of the way our own Galaxy is moving through space, which comes down unambiguously on the side of the just closed models, could scarcely be dismissed as unreliable when the traditional techniques were sending us the same message, and, as we shall see shortly, by the end of the 1980s the new analysis had stood up to every test that could be thrown at it.

The attraction of Virgo

Before we can have any confidence in estimates of the distance scale of the Universe based on studies of the redshift-distance law, we really need to be sure we understand all of the Earth's motion through space. Our home planet, on which all our telescopes are based, is moving around the Sun; the Sun is moving around the Galaxy; and the Galaxy itself is moving with respect to other nearby galaxies. Although all galaxies – or, rather, all *clusters* of galaxies – are being moved apart from one another by the universal expansion, they can each have quite substantial 'peculiar' velocities of their own, as they orbit around one another. The nearby galaxy M31, for example, is actually moving *towards* us, at present, because the galaxies in the Local Group are held together as a group by gravity, and are not being steadily pulled away from one another by the universal expansion. On this scale, the local gravity overwhelms the expansion effect. Cosmologists need to know all such local effects, and subtract them out of their calculations, before they can be sure that they are left with the pure redshifts due to the expansion of the Universe. We need a fixed frame of reference, a stationary platform which only moves with the universal expansion. Without such a reference point, all of the redshift surveys are at least partly based on guesswork.

The problem has been highlighted by changing opinions on just how big an effect the Virgo Cluster (3 metres away on the 'aspirin scale' discussed in Chapter 3) has on the motion of our Galaxy and the Local Group. We certainly are moving away from the Virgo Cluster, relatively speaking; the redshift shows that. But we would expect that the gravitational influence of all the matter in the Virgo Cluster would be holding us back a little, impeding our relative recession from the cluster caused by the expansion of the Universe. Confusingly, astronomers sometimes refer to this virgocentric pull in terms of a velocity of the Milky Way towards the Virgo Cluster as 'infall'; what they mean is that we are receding from the Virgo Cluster that much slower than we might expect from the Hubble law alone. But how much slower?

We can get some idea of what is going on by looking at the redshifts and distances of the galaxies in our immediate vicinity, where some astronomical techniques still provide an indication of distance independent of the redshift. If we look out in opposite directions in space and find two galaxies at roughly the same distance from us, but one of them has a slightly larger redshift than the other, we know that the excess redshift must be due to the peculiar motion of the Milky Way, or of one or both of the other two galaxies. If we look at enough galaxies in this way, we can hope that all the odd motions of the other galaxies cancel out, and any consistent tendency for redshifts on one part of the sky to be lower than those on the opposite part of the sky is a sign that our Galaxy has a peculiar velocity towards the low redshift region. This kind of technique has been applied to try to determine the size of our infall 'towards' the Virgo Cluster, but with only limited success. Different astronomers have come up with different estimates of the size of the effect, ranging from virtually no infall up to about 500 kilometres per second. A lot depends on which groups of galaxies are used to calibrate the motion of the Milky Way – and the astronomers who do the calculations are uncomfortably aware that their figures will still be distorted if all the galaxies being used in the calibration are themselves being held back by the Virgo Cluster in the same way. No peculiar motion of our Galaxy compared with the expansion of the Universe will show up if we try to measure the effect by comparing the motion of the Milky Way with the motion of a lot of other galaxies all streaming in the same direction we are.

Nevertheless, studies of the virgocentric flow are beginning to tell us something about the distribution of matter in the Universe.

The Virgo Cluster is just close enough for some estimates of its distance to be made using a variety of different secondary techniques. These involve some subtle astronomical reasoning, and do not all give the same 'answer' – indeed, the same technique applied by two different astronomers will often give two different values for the distance. The range of estimates is from about 16 Mpc to 22 Mpc, with 20 Mpc being a reasonable compromise between the extremes. Because of the uncertainty in the estimates of the peculiar velocities of the Milky Way itself and of the galaxies in the Virgo Cluster, this cannot be used directly to determine H. Instead, by comparing the brightnesses of individual galaxies within the Virgo Cluster and super-novae within those galaxies with their counterparts in a much more distant group of galaxies, the Coma Cluster, astronomers infer that the Coma Cluster is about six times further away than the Virgo Cluster – that is, some 120 Mpc. The Coma Cluster is so far away that it has a redshift corresponding to a velocity of 7,000 kilometres per second, far greater than the few hundred kilometres per second of the Milky Way's peculiar motion. So, at last, we have a more or less direct comparison of distance and redshift on a scale big enough to be sure that the peculiar motion of the Milky Way cannot be introducing an error of more than about 10 per cent. The long chain of reasoning yields a value for H of between 45 and 55 km/sec/Mpc. But this isn't the end of the Virgo story.

The strength of the pull which the Virgo Cluster exerts on the Milky Way depends on how much matter there is in the cluster. With this value of H, astronomers know how big the redshift 'ought' to be, and by comparing this with the measured redshift they infer that the effect of the attraction of Virgo is equivalent to a motion towards the Virgo Cluster at a speed of a little over 200 km/sec. The amount of matter needed in the Virgo Cluster to produce this effect is equivalent to a density of about one-tenth of the value required to close the Universe. Even if the 'infall' velocity is as high as 450 km/sec, we still only 'need' enough matter in the Virgo Cluster to produce a value for omega of about 0.25, if the same density of matter were spread uniformly throughout the Universe.

This would be a very powerful argument in favour of the Universe being open – *if* we could be sure that all of the matter in the Universe is distributed in the same way the bright galaxies are distributed (and assuming, of course, that the Virgo Cluster is typical of the Universe at large). But if there is any independent evidence of the Universe being closed, then what the attraction of Virgo is telling us is that not only is most of the matter in the Universe not in the form of bright stars, it isn't even distributed throughout the Universe in the same way that bright stars and galaxies we can see are distributed. What we need is a way to measure the distribution of matter over even greater volumes of space, looking at radiation in wavebands different from those of visible light on which astronomers have depended for so long. Twenty years ago, it would have been an astronomical pipe-dream. In 1986, it became reality.

Microwaves and the movement of the Milky Way

The technique of measuring the peculiar motion of our Galaxy through space – independent of the expansion of space itself – works best, in principle, using the redshifts of more distant galaxies. But the further away galaxies are, the harder it is to estimate their distances, and the less confidence we can have in the accuracy of the calculation. Nevertheless, back in 1976 Vera Rubin and her colleagues, of the Carnegie Institution of Washington, tried to extend the technique by comparing the motion of our Galaxy against the 'frame of reference' provided by a spherical shell of distant spiral galaxies. These galaxies are all at a distance of about 100 Megaparsecs from us, assuming that the Hubble parameter really is close to 50 in the usual units. They surround us in the way that the skin of an apple surrounds a pip at its centre, and they are so far away that it is reasonable to expect that all their own little, peculiar motions will average out, and that taken together they provide a reference frame moving only with the expanding Universe. Rubin's calculations showed that, relative to these distant galaxies, our Milky Way (and the Local Group) is moving through space with a velocity much bigger than anyone had expected – 600 kilometres a second, over and above our motion as part of the universal expansion. The discovery was so surprising, and the velocity revealed by the technique so large, that most astronomers simply refused to accept it. They could just about cope with an 'infall' of 200 or 300 km/sec towards the Virgo Cluster, where they could see evidence, in the form of bright galaxies, of the matter doing the pulling. But a velocity of 600 km/sec, in the direction of nothing in particular on the night sky, where there was no bright cluster of galaxies visible? Ridiculous!

Ten years later, the notion no longer seemed so ridiculous, and Rubin was vindicated. Two new pieces of evidence combined to change the opinion of the astronomers.

The first insight came from studies of the microwave background radiation, the leftover hiss of radio noise from the Big Bang itself. This radiation has filled the Universe since very shortly after the moment of creation, but it has not been affected by the material content of the Universe since electrons combined with the nuclei created in the fireball to make electrically neutral atoms. This kind of radiation can only interact with free charged particles; but within a million years of the moment of creation all of the positively charged protons and negatively charged electrons were locked up in neutral atoms of hydrogen and helium. Ever since, the background radiation has simply expanded with the Universe, cooling and weakening as it is redshifted to longer and longer wavelengths, but never being disturbed by matter. The background radiation ought to provide the best available frame of reference in the expanding Universe, an ideal basis against which to compare our own peculiar motion. And it does.

As observations of the background radiation have improved over the past thirty years, astronomers have been able to go beyond merely noting its existence and taking its temperature (observations that were instrumental in establishing the Big Bang description of the Universe) and have mapped the strength of the radiation over almost the entire sky, at many different wavelengths, using instruments that are now sensitive enough to measure small differences in the strength of the radiation – small temperature differences – from different parts of the sky. The techniques involve observations from the ground, from high-flying aircraft, from balloons which carry instruments above the bulk of the atmosphere, and from satellites in orbit around the Earth. By the middle of the 1980s they were showing, unambiguously, that there is a warm patch in the cosmic background, in a direction roughly at 45 degrees to the direction of the Virgo Cluster, and a cold patch in the opposite direction on the sky. The warm patch is equivalent to a region of blueshifted background radiation, where the wavelength has been shortened slightly because we are moving towards the incoming waves; the cold patch is a region of redshift, caused by our motion away from the incoming waves. The interpretation of the discovery is clear – we are indeed moving at a high velocity relative to the background radiation, and therefore relative to the overall expansion of the Universe. It is exactly the 600 km/sec velocity Rubin found ten years previously. Further observations made from space by the Soviet satellite Relict in the late 1980s and NASA's COBE (from 'COsmic microwave Background Explorer') in the 1990s have confirmed both the precise temperature of the background radiation (2.73 K), its uniformity, and our own motion relative to this smooth sea of radiation.[2]

At first, some astronomers thought that this motion might be due to the gravitational pull of a concentration of matter in a group of galaxies known as the Hydra-Centaurus Supercluster. If the Milky Way were being tugged one way by the Virgo Cluster and another way by the Hydra-Centaurus Supercluster, the overall effect could be to produce a movement towards a point in space roughly midway between the directions to the two masses. But that notion was squashed by a massive study by a team of astronomers from six different institutions around the world, from Herstmonceux in Sussex to Pasadena in California, who reported their investigation of the motion of 400 elliptical galaxies, spread evenly across the sky, to an international meeting in Hawaii in 1986. Using chains of argument like the one which seems to work so well for the Coma Cluster, they were able to work out distances and peculiar velocities for all these galaxies. They found that *all* of the nearby galaxies and groups of galaxies are being tugged through space in the same way as our Galaxy and the Local Group. The Virgo Cluster, the Hydra-Centaurus Supercluster, the Local Group and others are all moving, at 600 to 700 km/sec, towards a region *beyond* the Hydra-Centaurus Supercluster.

Where does the region involved in this streaming of galaxies end? And just how much matter do you need to pull so many galaxies so strongly? The best answers to these questions are provided by surveys of distant galaxies carried out by the Infrared Astronomy Satellite, IRAS, and, once again, reported in the mid-1980s.

Weighing the infrared evidence

All studies of the distribution of galaxies seen in visible light are handicapped by a phenomenon known as reddening. This has nothing to do with the redshift, but is a dimming and reddening of the light from distant objects caused by dust in the Milky Way itself – the effect is exactly equivalent to the way dust in the atmosphere of the Earth causes red sunsets. The dust of the Milky Way simply blocks out light from many regions of the sky, leaving astronomers with reasonably clear views of only parts of the northern and southern hemispheres of the sky, above the plane of the Milky Way. Light from faint galaxies (and that means, by and large, more distant galaxies) is affected worse, so the further out into the Universe you want to look the higher you have to raise your astronomical sights, to high latitudes in the northern or southern skies. Then there is the problem of comparing the northern galaxies with the southern ones. When they try to combine the limited observations they do have to provide a catalogue of galaxies covering as much of the sky as possible, the astronomers find it impossible to assess the brightnesses of northern and southern galaxies precisely in one definitive scale. Northern galaxies can only be studied by telescopes in the northern hemisphere; galaxies high in the southern sky are only visible to southern telescopes. Ideally, comparing the brightnesses of faint objects, measured at the limit of present-day techniques, to the precision required by these dynamic studies, requires that all the galaxies being studied are monitored with the same combination of telescope and instruments. But there is no way to use the same telescope and instruments to measure the brightness of every galaxy visible from the surface of the Earth; the telescopes are simply too unwieldy to move about.

IRAS solved both these problems, and others. Infrared light is scarcely affected by reddening caused by dust in the Galaxy, and the same instruments, in orbit around the Earth, were used to map the entire sky. IRAS could see galaxies in all directions except for a very narrow region of the sky across the Milky Way itself, and these galaxies could easily be distinguished from bright stars in our own Galaxy. The result was a survey of tens of thousands of galaxies at infrared wavelengths, covering almost the whole of the sky.

Some bright infrared galaxies have also been identified using optical telescopes, and their redshifts measured. By comparing the brightnesses of these objects in the infrared with those of other infrared galaxies that have not yet been studied optically, it seems that the IRAS survey extends out to distances at least twice as great as those of the galaxies studied by Rubin and her colleagues. But they are not distributed uniformly across the sky. On average, there are slightly more sources in same-sized areas of sky on one side of the sky than the other, and the direction picked out by the IRAS survey is almost exactly the direction in which we are moving compared with the cosmic background radiation. At last, astronomers can actually 'see' (using infrared detectors) evidence of a concentration of matter in the right direction to be producing the pull affecting the Local Group and other galaxies in our part of the Universe.

This isn't the end of the story. Michael Rowan-Robinson, then of Queen Mary College in London, was one of the researchers involved in analysing the IRAS data. He calculated how much matter there would have to be overall, distributed across the region of the Universe surveyed by IRAS in the same way that the IRAS galaxies themselves are distributed, in order for the extra concentration in the direction we are moving to produce a gravitational tug strong enough to give the Local Group a peculiar velocity of 600 km/sec. The answer he comes up with is exactly equivalent, within the inherent uncertainties of the approximations involved, to the density required to close the Universe. The simplest interpretation of the IRAS data is that omega is almost exactly one, and that the distribution of galaxies on the night sky picked out by the limitations of ground-based observations of *visible* light is *not* a good guide to the way matter is distributed across the Universe.

This is the most powerful piece of evidence concerning the nature of the Universe that studies of the dynamics of galaxies can yet provide. It is the first direct measurement of galaxy dynamics that gives a value for omega of one. These ideas came under intense scrutiny in the late 1980s and early 1990s, and they have stood up to every test. The basic picture is clear, and the observations mesh in very nicely with the growing conviction among theorists that $\Omega = 1$. Remember, too, that while it is always possible that we may find more matter, and more kinds of matter, in the Universe than we yet know about, there is no way that we can ever 'remove' the matter we have already found. That is an absolute bottom limit on the density parameter; estimates of omega must always go up as time passes, never down. Most astronomers are still cautious about making any dogmatic claims, but the evidence in favour of a closed Universe is better than it has ever been.

And yet, this flies in the face of what used to be the received wisdom, as recently as the early 1970s. Then, most astronomers were convinced that the Universe contained no more than about 20 per cent of the mass needed to make it closed. The mistake they had made was in assuming that the only matter in the Universe is the stuff we can see – bright stars and galaxies. It was only when theorists came up with a new model of how the Universe was born, a model that requires omega to be indistinguishably close to one, that they realized that there might be a lot more dark stuff out there as well.

Chapter 9

The Need for Inflation

Ironically, it had been the very success of the hot Big Bang model of the Universe that had lulled astronomers, for several decades, into thinking that bright stars and galaxies made up most of the matter in the Universe, even though some astronomers had suspected, since not long after the redshift–distance relation had been discovered, that there is more to the Universe than meets the eye. In the early 1930s, the Dutch astronomer Jan Oort was one of the pioneers who deduced the nature of the Milky Way Galaxy by studying the way the visible stars are moving. It was only at about that time that these measurements showed conclusively that the stars of the Milky Way are each in orbit around a centre quite distant from the Sun, moving in a way reminiscent of the way planets orbit around the Sun.

Our Solar System lies about two-thirds of the way out from the centre of this swirling system, in the galactic suburbs. In our neighbourhood, motions of nearby stars can be studied in some detail. They do not move perfectly in a single plane, but wobble up and down as they orbit the centre of the Galaxy, moving a little above and a little below the main plane of the Galaxy. The height to which a star moving at a certain velocity can climb out of the plane before it is pulled back down by the gravity of all the other material in the plane depends, of course, on the overall mass of the disk in the neighbourhood of that star. The more mass there is in the plane, the more tightly each individual star is held down to the plane by gravity. And by studying the distribution of the stars around the plane of the Milky Way, Oort showed that there must be three times as much matter in the solar neighbourhood as we can see in the form of bright stars.

Of course, he couldn't watch a single star moving up and down through the plane. These changes take thousands or millions of years. But the overall distribution of stars, the relative numbers at each distance above and below the centre of the plane, can be determined and compared with distributions deduced from the laws of orbital dynamics. These numbers give a reliable picture of the way gravity is constraining the movement of the stars. This kind of study shows that the stars are being held in place by several times more material than we can see in the bright stars themselves. Since the 1930s, about as much mass as there is locked up in the visible stars near our Sun has been identified as cold clouds of gas and dust spread between those stars, but that still makes up, together with the stars themselves, only two-thirds of the amount required to explain the local dynamics of the Galaxy.

Mass and light

This unseen, dark matter can be measured in terms of a number called the mass-to-light ratio, or M/L. This is defined to be 1 for our Sun – one solar mass of matter, in the form of a star, produces one solar luminosity of light. In the region of the Galaxy near the Solar System, Oort's figures tell us that M/L is about 3. That doesn't seem a very dramatic discovery, but at about the same time that Oort was finding evidence of dark matter ('missing mass') close to home in the Universe, Fritz Zwicky, a Swiss astronomer who made a lifelong study of distant galaxies, found evidence of dark matter on a much more impressive scale.

Zwicky was studying clusters of galaxies, groups containing several systems like our own Milky Way Galaxy which lie together in space. Our Galaxy is a member of a small cluster called the Local Group; it has only a handful of members. Some clusters contain hundreds of galaxies. Astronomers assume that these clusters are groups in which the galaxies are kept together by gravity, orbiting around each other but moving through space as a group, rather like a swarm of bees. But when Zwicky used the ubiquitous Doppler shift to measure the velocities of individual galaxies in one group, the Coma Cluster, he found that they were moving much too rapidly, relative to one another, to be held together by the gravitational pull of all the stars in all the galaxies of the cluster. It looked as if the flying galaxies ought to have moved apart, dissolving the cluster, long ago when the Universe was young. And he found the same thing when he looked at other clusters – they are all full of galaxies moving much too fast to be held together by the gravity of the stars we can see.

There are many uncertainties in this approach. The masses of the galaxies, for example, can only be estimated from their brightnesses, on the assumption that an average star in a distant galaxy is as bright as an average star in our Galaxy. The distances to the clusters themselves are uncertain, which also affects the argument. But the size of the effect that Zwicky found, and has since been confirmed by every similar study, is so big that it dwarfs any possible errors of this kind in the calculation. In round terms, the amount of matter needed to stop clusters of galaxies from evaporating away is so great that M/L rises to about 300 – there is *three hundred* times more dark matter in clusters of galaxies than there is in the form of bright stars. For comparison, the amount of matter required to close the Universe, distributed uniformly through space, is only three times this. If the Universe is closed, its overall M/L is about 1,000.

None of this worried astronomers very much in the 1930s, or even in the 40s, 50s and 60s. The nature of the expanding Universe, and even the fact that the Universe extended far beyond our own Milky Way Galaxy, were new ideas to astronomy over half a century ago, and there was ample room for speculation about how these observations might be resolved. On the small scale of Oort's discovery, it hardly seemed reasonable to believe that astronomers had found every kind of object that might exist in the Milky Way, and it was easy for astronomers to imagine that there might be many very faint stars ('brown dwarfs') or objects like large planets ('Jupiters') contributing plenty of mass, but very little light, to the Milky Way. These assumptions have, as we shall see, been borne out by recent discoveries. Zwicky's evidence was more puzzling, but in the absence of any evidence to the contrary the theorists could speculate that the space between the galaxies within clusters might be filled with a sea of gas, enough to hold the galaxies together by gravity. These assumptions have not been borne out by later observations, but the pioneers weren't to know that. It was only when the Big Bang theory became established as a good description of the real Universe that the puzzle of the missing mass came to the forefront of astronomy. For what had seemed in the 1940s to be a failure of the Big Bang model was, in fact, providing such a profound insight into the nature of the Universe that since the 1960s it has been able to tell astronomers, very precisely, how much matter there 'ought' to be in all the stars and galaxies – at least, in the form of everyday atomic matter.

Baryons and the Universe

Our everyday world is made of atoms. These come in many varieties, or elements – hydrogen and oxygen (sometimes combined to make up molecules of water), carbon, iron and the rest. Both the physical world and life itself depend on the interplay of atoms, combining in different ways and interacting to produce substances as diverse as the DNA which carries the genetic code in our cells and the gold that we value for its pretty colour and scarcity. But where do the atoms come from, and why is gold scarce on Earth, while water is abundant? These seemingly philosophical questions can be answered in great detail by astronomers, using the Big Bang model of the Universe.

Our home planet is far from being a typical part of the Universe. In terms of the atoms from which it is made, it is far from being a typical part even of our own Solar System. By far the bulk of the matter in the Solar System is concentrated in the Sun, round which all the planets orbit. The Sun alone contains as much matter as 333,400 planets like the Earth, and all the planets of the Solar System put together contain less than 450 Earth masses – less than 0.15 per cent of the mass of the Sun. The Sun is a much more typical representative of the Universe than the Earth is, and our Sun seems to be basically similar to the thousands of millions of other stars that make up our Milky Way Galaxy, which is itself basically similar to the hundreds of millions of galaxies that make up the visible part of the Universe. The Sun and stars do *not* contain the same elements, in the same abundances, that we find on Earth.

Remember that astronomers can find out what stars are made of by looking at their light. Each type of atom – each element – produces its own characteristic pattern of lines in the spectra from the stars, and the relative strength of these lines shows the proportion of each element present. Using radio astronomy techniques, it is even possible to probe the composition of cool clouds of gas between the stars in this way, and by and large the picture that astronomers get is always the same, wherever they look in the Universe. The great majority of all the material in all the stars and clouds of all the galaxies is in the form of hydrogen, the simplest element of all. A significant amount of the material in stars (about 25 per cent) is in the form of helium, the next simplest atom, but only a few per cent of the material in any star is in the form of heavier elements like the carbon, oxygen, iron and the rest that are so important on Earth. In some stars, less than one hundredth of 1 per cent of all the material present is in the form of heavy elements; all the rest is hydrogen or helium. And the stars which contain least in the way of heavy elements are always the ones that seem, from other evidence, to be the oldest stars in the Galaxy.

All this must be telling us something profound about the nature of the Universe. Hydrogen, after all, is the simplest element – an atom of hydrogen consists of a single proton associated with a single electron. In the most common form of helium, two protons and two neutrons together make up the nucleus of the atom, with two electrons outside it. Most of the Universe is made of the simplest kinds of atoms. Inside a star, the electrons are stripped away from the nuclei, and lead a more independent existence, but it is still nuclei of hydrogen (protons) and of helium (also known as alpha particles) that make up most of the mass of the visible universe.[1]

When Fred Hoyle and his colleagues showed how the heavier elements could indeed be produced by nuclear fusion, in just about the right proportions, by reactions going on inside stars, they were also showing that the only nuclei that had to be produced by the Big Bang were those of hydrogen and helium, with just a trace of a few other elements, such as deuterium. I mentioned in Chapter 6 that the precise amount of these other elements produced in the Big Bang is a very sensitive guide to the density of matter in the Universe – or rather, to the density of *baryonic* matter. It works like this. When gas expands, it cools. When a hot fireball of radiation expands, it, too, cools. From one point of view, you can think of this in terms of photons acting as the 'particles' of the radiation gas, losing energy like the atoms of an expanding gas cloud. For radiation, however, the effect of this loss of energy is to change the wavelength of the radiation, redshifting it by an appropriate amount. Energy that was in the form of gamma or X-rays, or even more energetic forms, is progressively redshifted down through the ultraviolet and the visible spectrum, on into the infrared and then into the radio wavebands. The exact temperature of this kind of radiation can be related to the exact way its energy is distributed at different wavelengths, and to the value of the wavelength for which the radiation has a peak intensity. It is called blackbody radiation, and the way in which its energy is shared among the wavelengths is called the Planck distribution, in honour of Max Planck, one of the founders of the quantum theory.

We saw in Chapter 6 how, a few seconds after the moment of creation, with the temperature of the Universe about 10 billion K, the fireball was already laced with a trace of baryonic material, in the form of neutrons and protons. With electrons also present, the two kinds of nucleon were to some extent interchangeable, since a proton and electron can be forced to combine, at high energies, into a neutron, while a neutron left to its own devices will decay into a proton and an electron. The first process, combining protons and electrons to make neutrons, became less and less common as the Universe cooled, and by the time the temperature was down to about a billion K, when the Universe was roughly 3 minutes old, the balance had shifted to the point where there were 14 protons and just 2 neutrons in every 16 nucleons. At this temperature, an individual proton and an individual neutron can stick together to form a nucleus of deuterium, or heavy hydrogen. When the temperature was higher, deuterium nuclei could form, but were immediately smashed apart by high energy photons; at 'only' a billion K, however, the deuterium nuclei not only hold together themselves but promptly combine in pairs to make nuclei of helium-4, each containing two protons and two neutrons. Out of every 16 nucleons, 4 have gone into helium – 25 per cent. The rest are left as protons, to become hydrogen atoms as the Universe cools further and they capture electrons. The presence of some 25 per cent helium in the oldest stars is the strongest evidence, together with the background radiation itself, that the Big Bang model is a good description of how the Universe we live in came into being.

In fact, though, the nuclear fusion reactions are a little more complicated than I have indicated previously. Instead of two deuterium nuclei combining directly to make one nucleus of helium-4, it is more likely that they will interact to produce a helium-3 nucleus and a lone neutron. The 'spare' neutron can interact with another helium-3 nucleus, almost immediately carrying the reaction through to helium-4. And that is still not quite the end of the story, since although there are no stable nuclei which contain either five or six nucleons, the fusing together of nuclei of helium-3 and helium-4 can produce a little lithium-7 before the Universe has cooled to the point where no more fusion reactions occur. Within 4 minutes of the moment of creation, however, all of this activity is over, and the primordial element abundances have been decided.

If we could measure the abundances of all of these elements in the oldest stars, then we would know the mixture that came out of the Big Bang. Because the exact proportion of each element produced depends on, among other things, the density of the fireball in which the elements were made, that could, if the Universe were made only of baryons, immediately tell us whether the Universe is open or closed, and therefore its ultimate fate.

Of all the lightest elements, the production of helium-4 is least sensitive to the density. But the amount of helium-4 produced does depend critically on the rate at which the Universe was (and is) expanding, so that measurements of helium-4 abundances are still of vital importance. It is easily the most accurately measured of all the abundances of elements heavier than hydrogen, and those measurements just fit with the requirements of the standard model of the Big Bang. The best modern estimates give a proportion of about 23 to 25 per cent of all the material in old stars as helium-4. Deuterium is equally interesting in principle, but very difficult to measure in practice. Deuterium is not manufactured inside stars at all, according to the modern understanding of nuclear physics. At the temperatures inside stars deuterium is actually destroyed. So any measurement of the abundance of deuterium in the Universe today must give a figure lower than the actual abundance which came out of the Big Bang. The measurements are difficult, but such techniques as measuring the amount of deuterium in samples from meteorites, and spectroscopic studies of the clouds of Jupiter, indicate that for every hundred thousand atoms of hydrogen in the Universe there are only two or three atoms of deuterium. Astronomers believe that perhaps twice as much deuterium, five atoms for every hundred thousand atoms of hydrogen, came out of the Big Bang, and that the rest has been destroyed inside stars.

Spectroscopic studies of the light from stars suggest that helium-3 was produced in roughly equal amounts to deuterium in the Big Bang, and lithium-7 is even rarer, with perhaps five atoms of lithium-7 having been produced in the Big Bang for every 10 billion atoms of hydrogen. All of these values for the element abundances can be explained in terms of the standard model of the Big Bang – provided that the total density of baryons in the Universe is significantly less than one-tenth of the critical value needed to make the Universe closed.

In its simplest form, the argument can best be seen in terms of deuterium. Deuterium nuclei in the early Universe are very likely to collide with one another and produce helium nuclei. When the density of nuclei is low, when they are few and far between, there will be relatively few collisions and so proportionately more deuterium will survive to be detected today. When the density of nuclei is high, there will be relatively more collisions, and fewer deuterium nuclei left for us to see. Even five deuterium atoms for every hundred thousand hydrogen atoms represents a high figure, on this basis. The present measurements of deuterium abundance set a very strong limit on the density of baryons in the early Universe, and therefore in the Universe today, and this limit is far below the critical value.

When the calculations of Big Bang nucleosynthesis were first carried through in this form in the 1960s, they seemed to provide conclusive evidence that the Universe was open, and would expand forever. That remained the view of cosmologists into the 1970s, for it never occurred to anyone to suggest seriously that the Universe might be largely made up of other forms of matter than the ones we see around us. It was taken for granted that baryons were the most important form of matter in the Universe. Indeed, not just baryons, but visible baryons, in the form of bright stars and galaxies. As recently as 1981, when I discussed the problem with John Huchra, of the Smithsonian

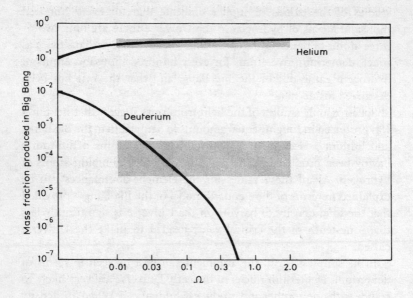

Figure 9.1 The Missing Mass
Standard models of the Big Bang set limits on how much helium
and deuterium can have been produced, as a fraction of the
amount of hydrogen, assuming most of the mass of the Universe
is in the form of baryons. The shaded boxes show the range of
options that is in agreement with observations of the Universe
today. The evidence tells us that there is no more baryonic
matter in the Universe than about one-tenth (0.1) of the amount
required to make the Universe closed, with $\Omega = 1$. Yet other
evidence shows that the Universe is very flat, and contains much
more matter than this. The missing mass cannot be baryons;
what is it?

Institution Observatory, he commented 'from a philosophical point
of view, an optical observer would quickly lose interest in observational
cosmology if the Universe was dominated by things he couldn't
see.' Well, Huchra's philosophical objections to the existence of
large amounts of dark matter in the Universe seem to have been
overruled – and the observers don't seem to have given up their trade,

but are busy studying the dynamics of the visible galaxies in order to probe, indirectly, the distribution of the dark matter in the Universe: dark matter that is not even baryonic. The supposition that the Universe is dominated by visible stars and galaxies now seems to have been based on nothing more than a kind of baryon-chauvinism. Just as the fact that elements such as carbon, oxygen and iron are common on Earth does not mean that they dominate the composition of stars and galaxies at large, so the fact that stars are composed chiefly of the baryons does not necessarily mean that the Universe at large is composed chiefly of the baryons. Even the figures for mass to light ratios were suggesting otherwise, before the new impetus to think seriously about non-baryonic matter came from the theorists in the 1980s.

Translating the constraints of baryon density set by the measured abundances of the lightest elements into these terms, M/L *must* be less than 72 for the whole Universe, if most of the mass of the Universe is in the form of baryons.[2] But we already know that for clusters of galaxies M/L is much bigger than this, 300 or so. The natural conclusion is that there is a great deal of mass 'out there' that is not in the form of baryons – and that is the conclusion that cosmologists have been forced to, initially reluctantly but with growing enthusiasm, in the past two decades or so. They now seem ready to embrace the idea wholeheartedly. After all, if there must be *some* non-baryonic matter, and there may well be a great deal of it, there might even be enough to push the value of omega close to one, where the best modern cosmological theories say it ought to be.

The problems of the Big Bang

In spite of its breathtaking success in explaining the broad features of the expanding Universe, there are several significant flaws with the standard model as outlined in Chapter 7. First, there is a difficulty known as the 'horizon' problem. It arises because the Universe looks the same in all directions. Even if we look at the distribution of galaxies and clusters of galaxies on the sky, the Universe seems fairly uniform; but a true indication of the uniformity of the Universe comes from studies of the background radiation, which is isotropic (the same in all directions) to better than one part in two thousand. How does radiation coming from one part of the sky 'know' how strong it must be in order to match so precisely the radiation coming from the opposite part of the sky – and indeed from all points in between? The observations we make today are the first contact we have had with radiation from these opposite 'sides' of the Universe, and these regions can never have been in contact with one another, according to the standard model, since they have always been further apart than the distance light can travel during the age of the Universe, at every epoch of universal history to date. If the 3 K radiation was emitted 500,000 years after $t = 0$, as the standard model implies, then only regions of sky that are less than two degrees of arc across, as viewed from Earth, can have been in contact with each other at the time. So the background radiation 'ought' to be patchy, with a grainy structure on a scale of two degrees or so. It looks as if the Universe was born out of the fireball era in a perfectly smooth state, with exactly the same energy density (the same temperature) 'built in', even in regions that were too far apart for any signal, restricted to travelling at the speed of light, ever to have passed between them. But what built this uniformity of temperature into the Big Bang?

The horizon problem leads directly to the second problem, the existence of galaxies. If the Universe was created in such a perfectly smooth state, how could lumps the size of galaxies ever have formed? In a perfectly smooth Universe, expanding uniformly, every particle of matter would be carried ever further apart from every other particle, and there would be no seeds which could grow by gravitational attraction into bigger concentrations of matter. You don't need very big seeds. Just irregularities corresponding to regions with density 0.01 per cent greater than the average, at a time 500,000 years after $t = 0$. But how could even such modest inhomogeneities be formed and emerge from a perfectly smooth Big Bang?

But the problem which triggered the new wave of research which led to a new theory of the evolution of the Universe at times prior to 10^{-30} second is called the 'flatness' problem. It comes back to old-fashioned studies of galaxies, redshifts and the expansion of the Universe, and it goes like this. We can calculate the expansion rate of the Universe, and we can estimate the amount of matter it contains –

or rather, the *density* of matter, which is what is important – by counting the numbers of galaxies. Einstein's equations allow for the possibility that the Universe may be open, and destined to expand forever, or closed, and fated to collapse back into a fireball. Or, just possibly, it might be flat, balanced on a gravitational knife edge between the two possibilities, as I discussed in Chapter 5.

However the argument described in Chapter 8 turns out, the observations show beyond doubt that the actual density of the Universe is definitely in the range from 0.02 to 10 times the critical density needed to make it closed. As cosmologists have argued about which side of the critical value the density might lie, they slowly came to realize that the fact that there is scope for such arguments at all is one of the most remarkable features of the Universe. Why shouldn't the relevant parameter be 10^{-4} of the critical value, or a million times bigger than the density needed to close the Universe, or some other figure so different from 1 that it would be obvious from observations which side of the line it lay? The Universe today is actually very close to the most unlikely state of all, absolute flatness. And that means it must have been born in an even flatter state, as Dicke and Peebles, two of the Princeton astronomers involved in the discovery of the 3 K background radiation, pointed out in 1979.

Finding the Universe in a state of even approximate flatness today is even less likely than finding a perfectly sharpened pencil balancing perfectly on its point for millions of years. For, as Dicke and Peebles pointed out, any deviation of the Universe from flatness in the Big Bang will have grown, and grown markedly, as the Universe has expanded and aged. Like the pencil balanced on its point and given the tiniest of nudges, the Universe soon shifts away from perfect flatness. The state of perfect balance is in equilibrium, but it is an unstable equilibrium, and any deviation from perfection spells disaster for the balance. We can wind the clock back, in our imaginations, to calculate how flat the Universe must have been during the fireball era for it to have a density still so close to the critical value today. Dicke and Peebles, and others since, have done the calculations for us. If the density of the Universe is now one-tenth of the amount needed to make it just closed, a figure that most astronomers would agree is a fair guide on the basis of the visible galaxies and calculations of baryon production in the Big Bang, then that means that 1 second after the moment of creation the density of the Universe was equal to the critical value to within one part in 10^{15}. And if we go back even further, at 10^{-35} second, the density must have been only one part in 10^{49} less than the critical value:[3] This is hardly likely to be a chance occurrence, and must mean that the laws of physics somehow require the Universe to be born out of the Big Bang in a state of extreme flatness.

In the spring of 1979, Dicke gave a talk at Cornell University, in which he discussed the flatness problem and pointed out how closely the Universe is balanced between runaway expansion and violent recollapse. One of the members of his audience was a young researcher called Alan Guth, a somewhat reluctant cosmologist who had been dragged into the subject by his friend and colleague at Cornell, Henry Tye. Guth was later to recall that he began working on cosmology only because he was pressured into it by Tye, who 'had to do a lot of arm-twisting, because at that time I very strongly believed that cosmology was the kind of field in which a person could say anything he wanted, and no one could ever prove him wrong.'[4]

In spite of this initial distrust of the subject, Guth found the ideas fascinating, and was slowly drawn deeper into cosmological studies. It took some months for the ideas presented by Dicke that spring day at Cornell to mingle with other ideas Guth was picking up from cosmology and particle physics – his original speciality – but by December 1979 they had begun to jell in his mind. In the course of the afternoon, evening and part of the night of 6 December, Guth put the pieces together in the first major new theoretical contribution to the cosmology of the early Universe since Gamow's work more than thirty years before. He introduced a completely new concept into cosmology, taking a step forward as profound as the idea of the cosmic egg itself.

The ultimate free lunch

Alan Guth followed a conventional route to become a research physicist in the 1970s. He was born in New Brunswick, New Jersey, in 1947, attended high school in Highland Park, New Jersey, where his family had moved when he was three years old, and entered MIT as a freshman to study physics in 1964. His Bachelor's and Master's degrees in physics were followed by a PhD awarded in 1972 - and then he ran up against the problems of finding a secure research appointment in physics. I know only too well what those problems were like in the early 1970s, since I completed my PhD work in 1971. There were plenty of young physicists with fresh PhDs around in those days, most of them much brighter than me, and very few secure jobs available. The only way to make the grade was to chase around from one short-term appointment to another, hoping one day to achieve the coveted position of a tenured scientist on a long-term appointment. I never even got on this treadmill, but quit research to take up writing instead; Guth, obviously a more persistent (and talented) character than I am, worked first at Princeton University, before moving on to Columbia University and then Cornell University, where he heard Dicke talk about the flatness problem in the spring of 1979. In October 1979 he moved on again, this time on leave from Cornell, to spend almost a year at the Stanford Linear Accelerator Center. It was there that the seed planted by Dicke's talk began to grow, fed by the input of all the pieces of information about cosmology and particle physics that Guth was taking in as the months went past.

Guth recalls the exact day that all of the ideas, gathering in his head, suddenly seemed to come together in one fell swoop. It was Thursday, 6 December 1979, and the breakthrough started with an afternoon spent in conversation with a visitor from Harvard, Sidney Coleman, in which they discussed the latest ideas. By evening, Guth was aware of something trying to crystallize in his mind. At home, after dinner, he settled down with his current notebook, and tried to give mathematical form to the ideas buzzing around in his head. He worked late into the night – long after his wife had gone to bed – filling page after page of the notebook with neat, meticulous calculations. When I was researching this book, Guth gave me photocopies of several of the pages from this and a later notebook. Towards the end of the section begun on 6 December 1979 and continued long after midnight into the small hours of 7 December, there is a short, five-line passage at the top of a page, carefully boxed around with a double line and headed in capitals SPECTACULAR

REALIZATION. It reads 'This kind of supercooling can explain why the universe today is so incredibly flat – and therefore resolve the fine-tuning paradox pointed out by Bob Dicke in his Einstein day lectures.' Guth had found the basis of the description of the very early Universe that is now known as inflation – he gave the model that name very soon after making the discovery. A couple of weeks later, Guth, by no means an expert in cosmology yet, learned for the first time, in a conversation with SLAC physicist Marvin Weinstein, that cosmologists were also puzzled about the large-scale uniformity of the Universe – the horizon problem – and realized that his new model could explain that too. Even though continued work on the model showed it to be flawed in at least one respect, he went ahead and published the paper in the hope, stated explicitly in the paper,[5] that someone else would find a solution to the flaw in it, and carry the idea forward. And, in the wake of his work on inflation, Guth became first a Visiting Associate Professor and then, in June 1981, Full Associate Professor in the Department of Physics at MIT – the post he still holds.

What was the model, flawed or not, that had made him so excited and was to revolutionize cosmology in the 1980s? It jumps off from the idea that the Universe may have experienced a change known as a phase transition, from an energetic to a less energetic state, during the first split second of creation, *before* the point at which the story of the Big Bang in terms of the standard model begins. Guth gave the name 'false vacuum' to the initial, high-energy state, and 'true vacuum' for the state with *lower* energy density corresponding to the way the Universe is today. He didn't just pull this idea out of a hat, however; there are sound theoretical reasons, to do with the way particles interact at very high energies, which make this notion of a phase transition plausible. But the details of the particle physics need not concern us here.

A useful analogy is with a marble in a large, smooth bowl. If the marble has a lot of energy – if it is rolled into the bowl, around the rim, quite fast – it can orbit around the bowl high up near the rim, like a stunt motorcyclist on the 'wall of death'. The position of the minimum, or the depth of the bowl, has no effect on its behaviour. But as its energy decreases, the marble slows and settles down into

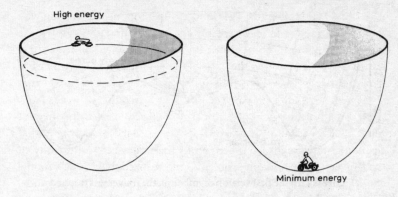

High energy

Minimum energy

Figure 9.2 The energy of the Universe can be likened to a wall-of-death motorcycle rider. When there is plenty of energy available, the rider can circle as high up the wall as he likes – this is equivalent to the creation of the Universe in a state of high energy density. When there is less energy, the motorcycle inevitably sinks down into the minimum. But in the case of the Universe, there may have been more than one minimum to choose from as it cooled and the energy density decreased.

the bowl, eventually coming to rest at the bottom, in a state of minimum energy. Now, the position of the minimum is all important. Something similar could have happened to the Universe as it cooled from 10^{-43} second to 10^{-35} second (when the temperature of the Universe was about 10^{27} K). It gradually settled down into a state of minimum energy. But which one? Suppose it settled into a state corresponding to the false vacuum, merely a *local* minimum, like the dip inside the crater of a volcano (Figure 9.3).

Such a state is likened to the way water can be supercooled, cooled below 0 °C without freezing. As the cooling continues, the water will eventually freeze, quite suddenly, and give up its latent heat of fusion in the process. Below 0 °C, ice is a more stable state with lower energy, but the transition into the more stable state, with accompanying release of latent heat energy, does not always happen at 0 °C. Guth said that the same sort of thing could happen to the vacuum of the very early Universe. While the Universe continues to cool below 10^{27} K,

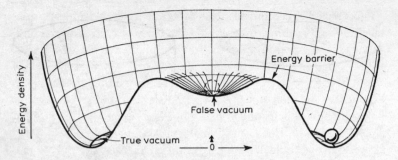

Figure 9.3 In the first version of inflation, the universe is trapped in the 'false vacuum' state until it tunnels through the energy barrier, like the alpha particle escaping from a nucleus, and settles into the true vacuum. Energy released in the process drives the rapid 'inflation' of the universe.

it stays for a time in the false vacuum state, like supercooled water staying liquid below its freezing point. The true minimum energy state is only reached when the Universe finds a way to tunnel through the barriers, and reach the deeper minimum of the true vacuum state.

The analogy with the volcano is a familiar one, of course. It is like the one used in describing alpha particles trapped in an atomic nucleus. Just as the alpha particle can tunnel out of its captivity with the aid of the uncertainty principle, so the early Universe can tunnel out of the false vacuum state with the aid of the uncertainty principle. But while the Universe is trapped in the false vacuum, or supercooled, state, its enormous locked-up energy density contributes, Guth found to his surprise, a huge outward push, making it expand far faster than in the standard model. The effect, for a short time, is like that of a cosmological constant far more powerful than any that Einstein imagined. As a result, the Universe expands exponentially, doubling its size repeatedly with every 10^{-34} second that passes. This may sound a modest expansion rate. But a doubling *every* 10^{-34} second means that in 10^{-33} second the region undergoes ten doublings and increases in size by a factor of 2^{10}, and in 10^{-32} second it increases in size 2^{100} times.[6] In far less than the blink of an eye, a region 10^{-36} times smaller than a proton itself can be inflated – hence the name for the model

– into a region 10 centimetres across, the size of a grapefruit. Inflation takes the vastly submicroscopic world of the very early Universe and very suddenly brings it up to the sort of dimensions we are familiar with.

But once the Universe tunnels into the true vacuum state, the rapid exponential inflation stops. The energy goes into the production of huge numbers of pairs of particles, which are reheated in the process almost to 10^{27} K. Virtually *all* of the matter and energy in the Universe as we know it could have been produced in this way out of the inflation. This possibility arises because the gravitational energy of the Universe is negative, and it is more negative the bigger the Universe is. As far as it is possible to talk about the way in which energy is conserved during the inflation, Guth says, the matter energy of the Universe increases, but the gravitational energy of the Universe gets more and more negative, so that the two together almost precisely cancel out – as Guth puts it, the Universe is 'the ultimate free lunch'. As the expansion of the Universe slows to the regular pace of the standard model of the Big Bang, the rest follows as in the standard model described in Chapter 7. The inflation itself is over, still much less than 10^{-30} second after $t = 0$; but it has solved the horizon problem *and* the flatness problem.

The solution to the horizon problem is simply that the regions on 'opposite sides' of the Universe today *were*, after all, 'in contact' just after the moment of creation, before they were blasted apart from one another by inflation. The Universe we see is very uniform today because it has all been blown up from a tiny, tiny seed in which all of the energy was uniformly distributed. The solution to the flatness problem is only a little more subtle. When you blow up a balloon, the surface gets less and less curved the more you inflate it. It becomes a better and better approximation to a flat surface. The same thing happens to the curvature of spacetime as spacetime is expanded by inflation. Whatever curvature you start out with, by the time spacetime has expanded by a factor of 10^{50} it is indistinguishable from a flat universe. Any curved universe becomes, as far as any of our observations could tell, a flat universe with density very close to the critical value by the time it has reached the size of a grapefruit.

By implication, however, there may indeed be *other* regions of

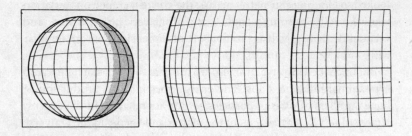

Figure 9.4 Inflation makes the universe flat, no matter what curvature it starts off with. In this example, the radius of the sphere is multiplied by three in each frame as we move to the right. The total 'flattening' is merely a factor of 9; in the inflationary scenarios, the universe is flattened by a factor of 10^{50}.

spacetime, beyond our visible Universe, that did not inflate at the same time or the same rate as our 'bubble' of the Universe. We may be living in a 'local' region (the entire observable Universe) of some much bigger meta-universe. But *might* those other regions be observable? This prospect, indeed, was the fatal flaw with the original inflation model.

As Guth realized, the transition from the false vacuum to the true vacuum would take place at random all over (or all through) spacetime. There would be many inflated bubbles of spacetime, each with their own particular values of the Higgs fields, and each with slightly different laws of physics, because the symmetry would break in a slightly different way in each bubble. There was no graceful way for the inflation to proceed smoothly in the original inflation model, and instead bubbles of inflation would be going off all through spacetime as different regions tunnelled out of the false vacuum in different ways. The bubbles would form clusters, rather like clusters of frog spawn in a pond, or like the structure of a sponge, and the boundaries between the bubbles would be very energetic, clearly detectable features. Our Universe looks nothing like this. So this prediction of Guth's original model is clearly wrong. The model is flawed. But the exponential expansion that was a feature of the first inflationary model looked so good, and solved so many problems, that many cosmologists wanted it to be true, in spite of its obvious flaws. The anguish of such a tantalizing, but flawed, possibility for resolving the prime puzzles of cosmology was summed up for me by Andrei Linde, then of the P. N. Lebedev Institute in Moscow, who took the next major step forward in 1981.

Linde was born in Moscow in 1948, and studied physics at Moscow University before moving on to research at the Lebedev Institute, under the direction, initially, of David Kirzhnitz. He was interested in the nature of the high-energy phase transitions involving the Higgs fields, and was intrigued to learn of Guth's work. He was delighted that the topic he had worked on with Kirzhnitz might be of cosmological significance, but frustrated by the difficulty of explaining how to achieve a smooth transition from the false vacuum to the true vacuum. Writing in March 1985, Linde told me that 'until the summer of 1981 I felt physically ill, since I could see no way to improve the situation and I could not believe that God could miss such a good possibility to simplify the work of creation of the Universe.' But in the summer of 1981, he hit upon a solution to the problem. In simple terms, what Linde suggested was that there is no deep well, like the inside of a volcano, associated with the false vacuum state, but instead there is a shallow plateau of energy, very slowly sloping off at the edges and down into the true vacuum state. A false vacuum on such a plateau would very gently and smoothly 'roll over' into the true vacuum, without the confusion of a whole series of local quantum transitions associated with tunnelling through the barrier. The result is a Universe which is smooth and uniform, one that has set like a mass of jelly instead of a handful of frog spawn. The idea became known as 'the new inflationary scenario', and although it took a little time to catch on, it gained wide popularity in the mid-1980s.

Linde's version of the theory was presented first at an international seminar in Moscow in October 1981. There, Stephen Hawking, of Cambridge University, responded with a 'disproof' of Linde's scenario. On second thoughts, however, he found the idea more beguiling, and both mentioned it during a seminar he gave at Philadelphia University and wrote a paper on the subject with one of his colleagues. Two Philadelphia researchers, Andreas Albrecht and Paul Steinhardt, were

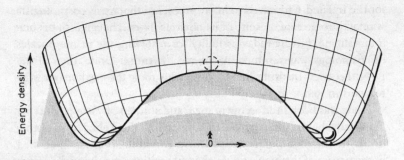

Figure 9.5 In the second variation on the inflationary theme, the Universe slowly rolls off the top of an energy plateau into the true vacuum. Once again, the equations suggest that the accompanying release of energy can drive a brief phase of extremely rapid, exponential expansion of the whole universe.

independently coming to much the same conclusions as Linde, and published their calculations in April 1982, referring in their paper to Linde's independent work; the 'new' inflationary hypothesis became respectable about that time (scientists always like to be told something twice before they believe it), and Linde's variation on Guth's theme was the front runner among inflationary models for the next couple of years.

The new inflationary hypothesis has been tinkered with further since 1982, in attempts to fine-tune it to produce a match with the observed features of the Universe. In one variation on the theme, the false vacuum state is surrounded by a *small* barrier, around the centre of the 'plateau', and tunnels out to form many bubbles which each begin their life on the plateau and then each get involved, separately, in a steady rollover into the true vacuum. The accelerated expansion of each bubble of spacetime continues as long as it is on the plateau, and each bubble can easily grow to the size of the observable Universe. At the time all this was happening, at the start of the phase transition, the distance that light could have travelled since $t = 0$ (the 'horizon distance') would have been about 10^{-24} cm. After inflation by a factor of 10^{50}, the same region would be 10^{26} centimetres across, all derived from an almost perfectly uniform speck of spacetime.[7] That volume of the inflated Universe would cover 100 million light years. But the size of the region of space within that vast volume that grew to become our visible Universe today would still have been only 10 centimetres, the size of a grapefruit, at the end of the inflation, just under 10^{-30} second after $t = 0$. Everything we can ever see or know about fits way, way down inside just one bubble of the inflationary Universe.

Many details of the inflationary scenario, or scenarios, remain to be worked out. The state of play today is rather like the state of play with standard Big Bang cosmology when Gamow was working on it in the 1940s. It is unrealistic to expect all of the answers to fall neatly into place just yet. Cosmologists now are working in two main directions. One line of attack seeks to explain how quantum fluctuations in the very early Universe can grow to become the seeds of galaxies a few minutes after the moment of creation. There is also the interesting question of just how important galaxies are in the Universe at all, gravitationally speaking. For, according to estimates of density based on counting the galaxies we can see, there is no more than 20 per cent of the matter needed to make the Universe flat available in the form of galaxies. And the calculations of the standard Big Bang which require there to be just 10 or 20 per cent as much matter in the form of baryons as is needed to close, or flatten, the Universe still hold good. So where is the other 80 per cent (or more) of the matter needed to make spacetime flat, as inflation says it must be? This is presumed to be non-baryonic matter. And, lo and behold, as I describe in detail in Chapter 10, when astronomers simulate the creation of galaxies in the Universe by modelling in a computer how gravity will pull clumps of matter together as the Universe ages and expands, they find that in order to match the observed clumpiness of matter they need a large amount of cold, dark matter, in the voids between superclusters of galaxies. The particles in the cold, dark matter each have masses of 1 GeV (about the mass of a proton) or more; they make up the bulk of the gravitational mass of the Universe, and they are not baryons. But the gravity of all these particles pulling together explains why the baryons that make up the visible stars and galaxies (and ourselves) are distributed the way we see them. But before I describe the search for this dark stuff in detail, there is a more recent variation on the inflationary theme, developed by Andrei Linde and a current front runner among theories of the very early Universe, which is worth exploring.

Primordial chaos and ultimate order

In spite of the conceptual breakthrough which Guth's work represented, even 'new inflation' only worked because the parameters are adjusted very delicately – 'fine tuning' which is permitted only because we don't really know how the false vacuum behaves, and which smacks of special pleading. The trick only works if the parameters are set up in just the right way in the first place, because we know the 'answers' we want to get out, and the inventors of the theory themselves accepted that this is implausible.

The later approach to inflation, pushing back to the moment of creation itself, tackles the puzzle of the appearance of the Universe out of a singularity – out of 'nothing at all' – according to the General Theory of Relativity. Linde, Guth and others realized, however, that this is not the complete story. According to the other great theory of twentieth-century physics, quantum mechanics, there is no such thing as nothing at all. Because of quantum uncertainty (the same quantum uncertainty that allows an alpha particle to tunnel out of a nucleus), what you and I might think of as 'empty space', or a perfect vacuum, is actually a seething mass of quantum fluctuations, in which pairs of particles constantly pop into existence, annihilate one another and vanish again, while spacetime itself disappears, on a scale 10^{-35} metre across, into a quantum foam. In quantum mechanics, there is no meaning to any length shorter than this 'Planck length', and there is no meaning to any interval of time shorter than the 'Planck time', 10^{-43} second. So little bubbles of energetic vacuum, 10^{-35} metre across and lasting for 10^{-43} second, are constantly appearing out of nothing at all and vanishing again.

Many of these quantum fluctuations will promptly vanish again. But not all. What happens if inflation sets in before the bubble disappears? On this scale, as Linde realized, it is meaningless to talk about the vacuum having any well-determined state, and it can only be described as a chaotic mess of interacting quantum fields. Because of this chaos and quantum uncertainty, at first the bubble of vacuum 'does not know' where the minimum energy state it 'belongs' in lies, and so it 'rolls down' into the true vacuum state only very slowly, with the tiny quantum bubble of nothing becoming an entire inflationary universe. This is alone enough to ensure a burst of inflationary expansion, without requiring either a phase transition or 'supercooling'.

Once the quantum field reaches its minimum value, the conditions which have been driving the inflation no longer exist. The field oscillates to and fro about the minimum, like a marble rolling to and fro as it settles into the bottom of a bowl, and in the process all of its energy is converted into pairs of particles. The marble slowly settles into the minimum energy state at the bottom of the bowl as friction converts its kinetic energy into heat; the oscillating field slowly settles into the minimum energy state as pair production converts the energy of its 'roll' into particles. As in the earlier inflationary scenarios, the Universe is reheated to nearly 10^{27} K. Then, it begins to cool rapidly as it expands in the more sedate fashion of the standard model. If this picture is correct, the order we see about us today was created by inflation out of primordial chaos; Linde calls his improved version of inflation 'chaotic inflation'.

The proliferation of variations on the inflationary theme is a sign both of the richness of the possibilities opened up by the new idea and of the seriousness with which cosmologists and particle physicists take the fundamental concepts. The basic idea itself now seems powerful enough, and well enough established, to accept, with as much confidence as Gamow accepted the idea of the Big Bang fifty years ago, that there was indeed a period of inflation early in the life of the Universe, which blew up everything we can see out of a seed no bigger than the Planck length, 10^{-43} centimetre, across. But it is far too early yet to say which, if any, of the different detailed scenarios now on offer will ultimately prove to be telling us about the real world. I'll leave the last words on the subject to the two pioneers, Guth and Linde. Linde is confident that progress is being made in the right direction. 'The old scenario is dead, the new scenario is old, and the chaotic scenario is in order,' he told me. Guth, who once thought of cosmology as a field where you could say anything and nobody could ever prove you wrong, has a slightly different view of things today.

'It now appears,' he says, 'that it is very easy to show that a cosmological scenario is wrong, and far more difficult than I had ever imagined to develop a totally consistent picture.'

Perhaps the inflationary models are not yet 'totally consistent'. But they have already given us one powerful and all important picture, an image of the moment of creation when the entire Universe as we know it was packed within the dimensions of the Planck length. And that makes it possible, as we shall see in Chapter 11, both to develop a mathematical description of the moment of creation itself and to gain new insight into the ultimate fate of the Universe. Before I reach my grand finale, however, I still have to explain what 90 per cent, or more, of the Universe is made of. Just what is the dark stuff that keeps the curvature of the cosmos so close to critical? And where is the missing mass?

Chapter 10

In Search of the Missing Mass

We now know that the elements are not really elementary. Atoms are made of baryons (specifically, protons and neutrons) and electrons. Because the mass of an electron is very much smaller than that of a nucleon, ordinary matter is often referred to as baryonic matter, and in calculating the density of the Universe on the basis of the way the lightest elements were synthesized in the Big Bang I didn't bother to add in the mass of all the electrons. This is reasonable enough, since although there is one electron in the Universe for each proton, the mass of each proton is roughly two thousand times the mass of each electron. In everyday units, the mass of an electron is 9×10^{-28} of a gram – that is, a decimal point followed by 27 zeroes and a 9. Everyday units are a bit impractical for measuring such small masses, and physicists prefer to use a unit called the electron Volt, or eV. Strictly speaking, this is a unit of energy. But mass and energy are interchangeable, since $E = mc^2$, and the c^2 term is usually taken as read. In these units, the mass of an electron is a little over half a million eV, written as 0.5 MeV, and the mass of a proton is 938.3 MeV while the mass of a neutron is 939.6 MeV.

These three tiny particles are the building blocks of everyday matter, all of the atoms in your body and in planet Earth, the material that makes up the Sun and all the bright stars we can see in the sky. But there is one more component of everyday matter that we have not yet been properly introduced to. It is called the neutrino.

In the 1930s, physicists realized that there must be a fourth kind of particle involved in nuclear reactions. When a neutron decays into a proton and an electron, for example, the energy of motion carried by the particles, as well as their mass energy, can be measured by suitable experiments. Whenever this was done, the experimenters found that the total energy being carried by the proton and the electron (mass energy + energy of motion) was less than the total energy carried by the original neutron. Something had to be carrying energy away. That 'something' was called a neutrino, by Wolfgang Pauli, as long ago as 1931. But neutrinos were not directly detected by any experiments until 1956.

The reason it took so long to detect neutrinos directly is that they interact only very weakly with ordinary matter. Neutrinos that are produced in nuclear reactions in the heart of the Sun, for example, pass through the whole thickness of the Sun itself more easily than light passes through a sheet of glass. Neutrinos are important in nuclear reactions, and anywhere that the density of matter or energy is very great, like the centre of a star or the Big Bang itself. But they pass through ordinary matter, such as planets and people, as if it wasn't there at all. But still, neutrinos are necessary to balance the books in nuclear reactions, and their existence has indeed been confirmed by experiments. They also complete a pleasing symmetry in the pattern of particles at the level important for nuclear physics.

Beyond the baryons

Protons and neutrons are both members of the family called baryons, and it seems to be a fundamental law of nature that the number of baryons in the Universe today stays the same. When a neutron decays, it doesn't just disappear in a puff of energy, or turn into a different kind of particle; it turns into a different baryon. But along the way it seems to create an electron, so obviously electrons are not conserved in the way that baryons are. The presence of the neutrino, however, restores the balance. The electron and the neutrino are together members of a family, called leptons. And just as the number of baryons in the Universe is conserved, so is the number of leptons.

You can picture the decay of a neutron in two ways. In the first, a neutrino (lepton) is absorbed by a neutron (baryon) and this then decays into a proton (baryon) and an electron (lepton). One lepton and one baryon are converted into a different lepton and a different baryon. From the other point of view, a neutron on its own can decay to produce a proton (conserving baryon number) and also an electron plus an *anti*neutrino. Just as electrons have antiparticle counterparts called positrons, so neutrinos have antiparticle counterparts. In adding up the total number of leptons in the Universe, you have to balance the books by subtracting the antiparticles from the particles. So by 'creating' a lepton (electron) and an antilepton (antineutrino) the neutron decay has maintained the balance of leptons in the Universe. This is an important realization, because nature does seem to favour this kind of symmetry. As physicists have probed deeper into the particle world over the past fifty years, developing mathematical rules which describe particle interactions ever more completely, they have found symmetry and balance at every level – for there are indeed levels, or at least one level, beyond the baryon.

All the evidence we have is that leptons are truly fundamental particles. There is no structure within an electron or a neutrino, and neither of them can ever be divided into smaller components. But baryons are different. In the 1960s, physicists established that the behaviour of baryons (and of other kinds of particle, which needn't bother us here) can best be explained if each proton and each neutron is made up of three more basic particles, called quarks. Just two types of quark are needed to account for the properties of nucleons. They have been given the arbitrary names 'up' and 'down', and a proton is best thought of as a tightly knit group made up of two up quarks and one down quark, while a neutron is made of two down quarks and one up quark. Other combinations of these two types of quark, including pairs of quarks, can, together with the lepton pair, explain all of the behaviour of everyday matter. Although individual quarks cannot exist in isolation, the up/down pair has many similarities in its basic properties to the electron/neutrino pair. Like the leptons, quarks have no internal structure and seem to be truly fundamental particles, and in the early 1970s it seemed that physics was on the brink of explaining the material world in terms of just four types of truly elementary particle. Since then, things have got a little more complicated, but even the complications seem to preserve the basic symmetry between quarks and leptons.

The complications can best be visualized as nature repeating her quark-lepton theme – not once, but twice. As long ago as 1936, physicists studying cosmic rays discovered a particle that seemed to

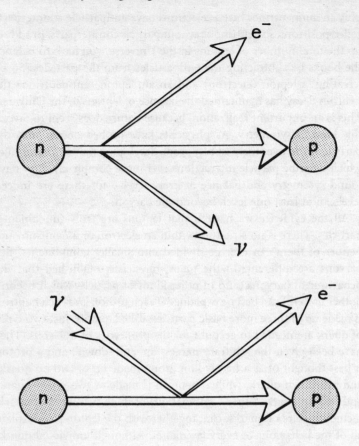

Figure 10.1 The Need for Neutrinos

When a neutron 'decays' into a proton and an electron, the equations can only be balanced by including another particle, the neutrino. Strictly speaking, a neutron decay produces a proton, an electron and an electron antineutrino; alternatively, an incoming neutrino colliding with the neutron is converted into an electron, while the neutron is converted into a proton.

be identical to the electron except that it has 200 times as much mass. It was called the muon, and often referred to as the 'heavy electron', but for forty years nobody had a clue what role it played in the particle world. Then, in the mid-1970s, the world of particle physics was rocked by the revolutionary discovery of new kinds of particle with larger masses than their 'first generation' counterparts.[1] These particles, and their behaviour, could only be explained in terms of the existence of two more quarks, heavier than the up/down pair, which were dubbed 'charm' and 'strange'. Just as the up/down quarks and the electron/ neutrino leptons formed a family, so did the charm/strange quarks and the muon and its own neutrino. The discovery of this 'new' generation of particles gave physicists a test-bed for their theories, which had been developed on the first generation, and the theories passed the tests with flying colours, predicting, or explaining, the properties of the new particles and their family relationships to one another.[2] Nobody knew why nature should have repeated herself, but the physicists were delighted at the opportunity to confirm and refine their models. Scarcely had they done so, however, when they had more to think about.

If one heavy electron can exist, why not more? With particle physics in turmoil in the mid-1970s, a team at Stanford University searched for evidence of a superheavy electron, and found it: a 'new' lepton, identical to the electron except for its mass, an impressive 2,000 MeV – an 'electron' that weighs twice as much as a proton! The new particle was given the name tau, and, as expected, turned out to be partnered by its own type of neutrino. There were now *three* generations of lepton, but only two generations of quark. The implications were obvious, and the search for a third generation of yet heavier quarks was on. Traces of one, called bottom, were found in 1977; evidence for the existence of the other, called top, has come from experiments at CERN in the mid-1980s. The quark–lepton symmetry is restored, those favoured theories of the physicists have been tested yet again and still come up trumps, and once again we have a balanced picture in which now six kinds of quark and six kinds of lepton are needed to account for everything in the material world, while only two of those quarks and two leptons are needed to account for all the matter in the everyday world. But where will it all end? If nature can indulge herself by allowing heavier copies of the two basic quarks and the two basic leptons, and can then make yet a third generation of particles, given enough energy, perhaps the process is endless. Why stop at three generations, or four, or five? Just when it seems we have got too deep into the particle swamp to have any hope of escape, however, cosmology comes to the rescue. There *are* good reasons why we should not expect the particle physicists to discover, or manufacture, any more generations of heavier quarks and leptons – and those reasons have to do with the influence of neutrinos on the way the lightest elements were synthesized in the Big Bang.

Neutrino cosmology

The 'extra' generations of particles do not exist in the Universe today, except where they are made in high-energy events. When they are made, they soon decay into the familiar particles of the everyday world. So there is no possibility that particles made of the heavier quarks (charm, strange, top and bottom) can provide a significant amount of extra mass in the Universe, over and above the mass of the baryons produced out of the Big Bang.[3] But the extra generations of particles can exist where there is enough energy, and there was certainly enough energy for them to have played a part in the processes going on in the Big Bang, just before the time, about three minutes after the moment of creation, when the baryons settled down into a mix of 75 per cent hydrogen, 25 per cent helium-4 and a trace of deuterium, helium-3 and lithium-7. The existence of the heavier quarks does not affect the subsequent story of the Universe at all, and for practical purposes we can ignore them. But the presence of three kinds of neutrino during the early stages of the Big Bang does turn out to be very significant in determining how fast the Universe expands, and therefore influences its ultimate fate.

The important point is the number of distinct varieties of light particle that exist. 'Light', in this context, means with a mass of less than about a thousand electron Volts, 1 keV. The electron neutrinos were originally thought to have no mass at all, and certainly fit this requirement; more recently, as we shall see, there have been suggestions that the electron neutrino may have a mass of a few eV, and that there may be enough of them to provide all the missing mass. This no longer looks the best bet in terms of dark matter candidates, but neutrinos are, indeed, very light. Such light particles are also sometimes referred to as 'relativistic' particles, because they emerge from the Big Bang with very large velocities, close to the speed of light,[4] which we know from relativity theory is the ultimate speed limit in the Universe. By analogy with the rapid motion of energetic particles in a hot gas, they are also referred to as 'hot' particles. In that terminology, neutrinos left over from the Big Bang which carry a trace of mass would represent the presence of hot, dark matter in the Universe.

But that is getting ahead of the story. For primordial nucleosynthesis, especially of helium-4, what matters is that the three kinds of neutrino are indeed distinct. This shows up in experiments here on Earth. When I said earlier that a neutrino can interact with a neutron to produce a proton and an electron, I should have specified that an *electron* neutrino was involved in the interaction. If a muon neutrino interacts with a neutron, then, as we might expect, the interaction produces a proton plus a muon. *How* the three flavours of neutrino 'know' which partner they belong to physics cannot, yet, tell us. But they do know, and the flavours are different. Armed with that information, cosmologists can set very precise limits on the amount of helium-4 that could have been produced in the Big Bang.

The amount of helium-4 produced by nucleosynthesis in the Big Bang does not depend critically on the density of the Universe. Roughly 25 per cent helium emerges from the Big Bang for a wide range of possible densities. But the amount of helium-4 produced does depend on how quickly the Universe is expanding at the time of nucleosynthesis, in such a way that *more* helium-4 is produced if the Universe is expanding faster. This is because if the Universe is expanding rapidly, more neutrons can be locked up in helium nuclei before they have time to decay into protons. If the Universe expands slowly, more neutrons decay before the temperature falls to the point where helium nuclei are stable. The rate at which the Universe is expanding at the time of nucleosynthesis depends on how many different kinds of relativistic particle there are in the Universe at that time. You can think of this as like a pressure forcing the Universe to expand – gas under pressure in a cylinder will force a piston to move outwards as the gas expands, and the more gas there is packed into the cylinder to start with the more pressure there will be and the faster the piston will move. The analogy isn't exact – in the early Universe, this particular 'pressure' does not involve the massive particles, such as protons, but only the light particles of the relativistic 'gas'. These include photons, electrons and positrons, and the three types of neutrino and their antineutrino counterparts. And the predictions for helium-4 abundances that come out of the calculations are astonishingly precise.

We know for sure, of course, that the photons and electron/positron pairs were present in the Big Bang. So, taking those as read, we can

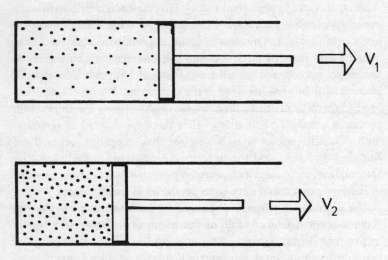

Figure 10.2 The Universal Pressure Cooker
The more tightly compressed a piston is, the harder the trapped gases push outwards. Let the piston go, and the pressure will make it move faster (v_2) than if the gas is less compressed (v_1). The speed with which the Universe expanded during the Big Bang depends on how many neutrinos (among other things) were doing the 'pushing'; the amount of helium manufactured in the Big Bang depends on how long the 'pressure cooker' regime lasted. By balancing the two requirements and measuring the proportion of helium in old stars today, astrophysicists deduce that there are certainly no more than four families of neutrinos, and that the three kinds already discovered are probably all there are to find.

express the predicted helium-4 abundances in terms of the number of neutrino flavours present. With only two types of neutrino (and their antineutrino counterparts) allowed for in the calculations, less than 23 per cent of the baryonic matter can be in the form of helium-4. With three types of neutrino, the proportion of helium-4 rises to 24 per cent, while with four neutrino flavours the predicted abundance is rather more than 25 per cent. In round terms, the addition of each extra flavour of neutrino increases the amount of helium compared with hydrogen by one percentage point. The latest measurements of helium abundance in the real Universe give a figure of 23 to 25 per cent, exactly in agreement with the prediction if there are three neutrino flavours only. The figures can just be stretched to include the range corresponding to four flavours, but no further; the best evidence from cosmology is that all the varieties of neutrino that exist in the Universe have already been identified.

The extraordinary power of this cosmological insight into the world of particle physics can be seen by looking at how difficult it has been for the particle physicists to decide, on the basis of their experiments carried out here and now on Earth, how many (or how few) neutrino flavours there ought to be. In the early 1980s, using various indirect arguments, and a lot of wishful thinking, the physicists could only say that there must be less than 737 varieties of neutrino. Over the next few years, they struggled to push this limit down, first to 44, then to 30 flavours. In the mid-1980s, high-energy experiments at CERN showed that the limit could be set as low as six or seven flavours. But it was only at the end of the 1980s that the CERN experimenters pinned the number of neutrino varieties down to precisely three – and cosmology certainly got there first. The beautiful way in which everything hangs together is persuasive evidence for both cosmologists and the particle physicists that the standard model of the Big Bang is correct, that there are three flavours of neutrino in the Universe, and that the amount of baryonic matter in the Universe is only enough for an omega of about 0.1. But the fact that *any* baryonic matter exists at all is, from one perspective, something of a puzzle. The Big Bang fireball was, after all, initially a sea of energy – radiation, revealing its presence today in the form of the cosmic background – which gave birth to baryonic matter as electromagnetic energy was converted into pairs of particles in line with Einstein's equation, $E = mc^2$, and the laws of quantum mechanics. Why didn't every particle created in this way pair up with its equivalent antiparticle and annihilate, leaving nothing but radiation?

Making matter

If the implications of Hubble's observations of the redshift and Einstein's General Theory of Relativity really can be pushed back that far, the quantum rules seem to be telling us that the Universe was 'created' at an 'age' of 10^{-43} second. The moment of creation itself is the subject of the next chapter. But in a Universe initially full of energy, with particles existing only as a result of pair production in which a matter particle, such as a quark, is accompanied by its antiparticle counterpart, how come there was any matter left over to form neutrons and protons at all? Why didn't all the quark–antiquark pairs (and electron–positron pairs) annihilate when the Universe cooled and no more pairs were being produced?

The explanation begins in 1956, with work by Chen Ning Yang and Tsung Dao Lee, two Chinese-American physicists. Lee, at Columbia University, and Yang, at Princeton, had met several years before, when both were at Berkeley, and kept in touch to discuss problems in particle physics. At that time, the number of known varieties of particles (sometimes referred to as 'the particle zoo') was rapidly increasing, and the behaviour of some of the newly discovered particles was baffling. One family, the kaons, did not seem to be acting as they should – their decay into stable particles such as electrons and neutrinos seemed to violate some of the rules of the particle physics game. Lee and Yang found a way round this, by proposing that in some interactions there is an occasional violation of a rule known as the conservation of parity. This rule holds that the laws of nature are unchanged if they are reflected in a mirror – that nature cannot tell left from right. But this was an assumption that had never actually been tested. Lee and Yang startled the physics community by pointing this out and suggesting that kaon decays violate parity conservation (P); within a few months experiments had been carried out (by another

Chinese physicist, Chien Shiung Wu, who had also been born in Shanghai, in 1912) that confirmed their prediction, and Lee and Yang shared the Nobel Prize in 1957, only a year after publishing their ideas. The speed with which the award was given indicates the dramatic impact of this overturning of a central idea in physics. It gave physicists a new way of looking at the particle world, and took the blinkers of an old dogma away from their eyes.

In 1964, two American physicists, James Cronin and Val Fitch, working at Princeton, extended this work to include another form of non-conservation. They showed in their experiments that kaon decay also, very occasionally, allows particles to change into antiparticles. The process breaks a conservation 'law' (really just dogma, based on 'common sense') known as charge conjugation, or C. Cronin and Fitch showed that C is sometimes violated on its own, and that a combination of C and P together, CP, can also be broken in some interactions. The changes in C do not always cancel out the changes in P. Cronin and Fitch duly received their Nobel Prize in 1980.[5] But this left physics with an interesting oddity. Apart from reflecting them in a mirror and swapping particles for antiparticles, the only other way to 'reverse' a particle interaction is to make time run backwards – or, if you prefer, to make a film of an interaction and then run the film backwards. Together with C and P invariance, the addition of time (T) invariance makes a rule known as CPT symmetry. In the early 1950s, physicists already knew, and many experiments have since confirmed, that if you were to do the equivalent of making a film of a particle interaction, swap all the particles for antiparticles, reverse left and right, *and* run the film backwards the end result would look exactly the same as the original. The laws of physics *are* unaffected by a complete CPT transformation, which is why it had seemed natural to assume that each component of CPT was itself 'symmetric' in the appropriate sense. But in fact each of the three components – C, P and T – can be 'broken' separately, provided that the combined CPT symmetry still holds in any particular case. The changes caused by C breaking, for example, may be cancelled out by the changes involving P breaking during the same interaction; but if CP invariance is broken, the only way to restore symmetry for the whole CPT system is if T is also broken, in just the right way to cancel out the CP breaking.

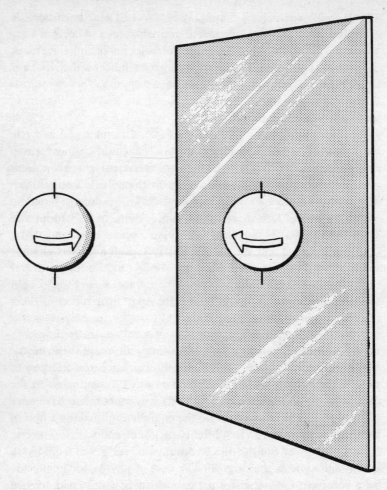

Figure 10.3 In the mirror world, the original motion of the sphere is restored if time runs backward. The mirror reflection and the flow of time are both asymmetric, but put together they produce symmetry – like the CPT symmetry, which is also made up of pieces that need not be symmetric themselves.

What this means in practical terms can be seen by looking again at the kaon, or specifically at one type of kaon, called K_L^o. This is a neutral particle that, like the photon, is its own antiparticle. It can decay in two important ways. One decay produces a negative pion, a positron and a neutrino. The other produces a positive pion, an electron, and an antineutrino. Don't worry about the pions – they are similar to photons, the 'particles' of light, and are not subject to the conservation laws. But do those conservation laws apply to leptons and baryons *all* the time? If the laws of physics were symmetric, then we would find that out of a large number of kaons half would follow each mode of decay, maintaining the balance between electrons and positrons (anti-electrons) in the Universe. In fact, the first decay mode is very slightly more common. As kaons decay, they increase the proportion of positrons in the Universe by a tiny amount.

This was the first chink in the symmetry of the laws of physics at this level. The Universe seemed to show a very slight preference for left-handedness, at least as far as electrons were concerned. But what about baryons? In the mid-1960s, nobody had a good theory of the key interaction involving baryons, known as the strong force, so no comparable calculations to those of Lee and Yang could be carried out for protons and neutrons. And the quark idea was scarcely more than a glint in the eyes of the theorists – while there was, and remains, no prospect at all of experiments on Earth to test theories of the strong interaction directly. Against that unlikely background, however, the brilliant Soviet physicist Andrei Sakharov set out, in 1967, the underlying principles that must apply to any process which could produce matter particles preferentially in the early Universe.[6] In order to come out of the Big Bang with an excess of matter over antimatter, Sakharov said, three conditions must be satisfied. First, there must be processes which produce baryons out of non-baryons. Second, these baryon interactions – or, at least, the ones that matter – must violate both C and CP conservation. Otherwise, even if baryons are made by some process, there will be an equal number of antibaryons made in the equivalent antiprocess, and eventually the particles and antiparticles will meet up and annihilate. And, third, the Universe must evolve from a state of thermal equilibrium into a state of disequilibrium – there must be a definite flow of time, so that CP processes together can be non-conserved, even though CPT remains conserved.

In physical terms, it is easy to understand this need for an 'arrow of time' in the Universe. If radiation and matter are in balance, at a uniform high temperature, then all the processes which turn radiation into particles and particles into radiation, or particles into other particles, are in balance, and proceed equally happily in either direction. In order for matter to be left over in the Universe today, the particles which ultimately decay in such a fashion that a residue of matter is left over must be produced in such a hot state of the Universe. But as long as the temperature is high, the reactions that make matter are being balanced by the reverse reactions which use up matter and turn it back into those primordial particles. It is only if the temperature of the Universe falls that the balance of the equations is tilted in one direction, in favour of the lower energy state – it so happens, in the direction of decays which produce a surplus of matter – and at the same time the fact that the Universe is evolving from a hot state to a less hot state provides a direction for time. You need a high energy state to start with, plus a move towards lower energies, in order to get matter in the Universe.

Sakharov was able to state these conditions on the basis of fundamental physical principles alone, without knowing what forces or particles would be involved in the creation of matter, nor at what precise temperature the balance of the equations would tilt in favour of matter. He was far ahead of his time, and his paper was published in Russian; it made no great impact in the late 1960s. The idea lay quietly for ten years. But it surfaced again in the work of a Japanese physicist, Motohiko Yoshimura, of Tohoku University, in 1978. Yoshimura had been working on grand unified theories, or GUTs, which attempt to explain all of the forces of nature in one mathematical package. The single most solid prediction of all the GUTs is that there ought to be unstable particles known as X bosons that have masses of about 10^{15} GeV – 10^{15} times the mass of the proton. Since X bosons are unstable, none of them would actually exist on Earth; but, according to theory, they would be produced out of pure energy anywhere where the temperature was high enough – such as in the Big Bang. To be precise, such temperatures existed during the first 10^{-35} second of the existence of the Universe, a time known as the GUT era. With no hope of ever detecting such massive particles in

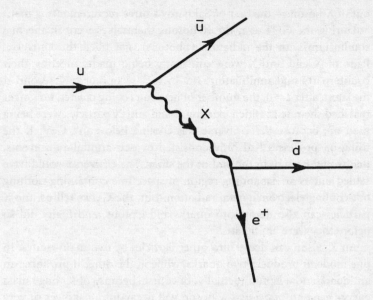

Figure 10.4 Interactions involving an X boson can make baryons out of non-baryons, or vice versa. Quarks and leptons are interchangeable when there is enough energy available for X bosons to exist.

accelerators on Earth, Yoshimura developed the idea that the X bosons might be revealed by their influence on the Universe during the GUT era, before 10^{-35} second, when they would have been the dominant constituent of matter. Following Yoshimura's lead – and only later rediscovering Sakharov's pioneering work and giving him due credit – theorists soon found that X bosons could produce an excess of baryons over antibaryons at the end of the GUT era, just as kaon decay can produce an excess of positrons over electrons today. The excess is small, but it is sufficient to do the job.

Of course, all these interactions really involve *quarks* and leptons, not protons as such, but the principle is the same. An up quark can change into an anti-up quark by emitting an X boson which decays into an anti-down quark and a positron. The GUTs specify a prescription for turning quarks into leptons, and also for creating quarks *and* leptons out of X bosons – the first of Sakharov's three requirements is met.

During the GUT era, the X bosons themselves were in thermal equilibrium with the radiation (photons) that filled the Universe. Pairs of X and anti-X were constantly being made, meeting their counterparts, and annihilating. But by 10^{-36} second or 10^{-35} second at the latest, after $t = 0$, the number of new pairs being created no longer matched the rate at which existing X and anti-X particles were being used up, because the Universe was cooling below 10^{15} GeV. If the using-up processes had only consisted of pair-annihilation events, that would have been the end of the story. The Universe would have settled out as an expanding region of spacetime containing nothing but cooling electromagnetic radiation. But, the GUTs tell us, the X particles can also decay into quarks and leptons, and many did so before they were annihilated.

An X boson can decay into other particles by two main routes. In one mode, it produces two quarks, while in the other it produces an antiquark and a lepton (actually, of course, because of its huge mass energy even an *individual* X boson will decay into a shower of very many pairs of quarks and quark–lepton pairs, but let's take things one step at a time). Decay of anti-X, similarly, will produce either pairs of antiquarks or pairs made up of a quark and an antilepton together. But because the decay processes violate the C and CP symmetries, they do not work at the same rate for anti-X as they do for X, and Sakharov's second requirement is met. In very approximate terms, when anti-X decays to produce a billion antiquarks, the equivalent amount of X decays to produce a billion and one quarks. And that is the source of all the matter in the Universe today. As the Universe expanded and cooled (meeting Sakharov's third criterion) one quark in a billion (1 in 10^9) failed to find an antiquark partner after the GUT era, and so never annihilated but instead stayed around to form the protons and neutrons that make up the bulk of the matter in the Universe today.[7] We owe our existence to a very tiny imbalance in the laws of physics, a preference for matter over antimatter in the decay of X bosons that amounts to no more than one extra quark for every billion antiquarks – an imbalance equivalent to one ten millionth of one per cent of all the matter that existed in the form of X and anti-X pairs in the GUT era.

The number 10^9 is not pulled out of the hat. It is the ratio of photons to baryons in the Universe today, calculated from the density of matter and the intensity of the cosmic background radiation. None of the GUTs is good enough to predict precisely what the imbalance between matter and antimatter 'ought' to have been at the end of the GUT era, and estimates fall into a rather broad range for the photon/baryon ratio, anything from 10^4 to 10^{13}. At least the 'required' value does lie somewhere in the middle of that range! But what matters most, of course, is simply that GUTs provide in principle a mechanism to make the matter in the Universe. It was this, as much as anything else, that encouraged cosmologists to embrace the new physics of high energies so warmly in the late 1970s and early 1980s. Perhaps, they thought, high energy physics could explain away some of the other remaining deep mysteries surrounding the standard model of cosmology, such as the horizon and flatness problems. As we have seen, they were proved right – but at the cost of having to find at least ten times more matter in the Universe than we can see in the form of bright stars and galaxies, ten times more matter, in fact, than can possibly exist in the form of baryons!

The bulk of the dark matter required to flatten the Universe (in the favoured cosmological models) or even to hold large clusters of galaxies together (based on dynamic evidence) is not baryonic. So what is it?

Cold, dark matter

The limit set by cosmology and the helium abundance on the number of neutrino flavours is actually even more stringent than I have so far spelled out. The fact that the actual helium abundance in the Universe is about 25 per cent (this is the 'primordial' abundance, after allowance has been made for helium manufactured inside stars since the Big Bang) is consistent with there being only five kinds of relativistic particle at all playing any significant part in the expansion of the cosmic fireball at the time of nucleosynthesis. These are the photons, electrons/positrons and the three flavours of neutrino. The particle physicists tell us that there may be other kinds of particle that play a part in particle interactions at high energies today. Although they have not yet detected any of these particles directly, some of them seem to be needed in order to round out the symmetry of the equations in the most successful particle physics theories, and the theorists have happily given these hypothetical particles names, such as photino and gravitino. Cosmologists refer to them collectively, perhaps slightly tongue-in-cheek, as 'inos'.

Although the theories that predict, or allow, the existence of inos are regarded as the best theories of particle physics we have, they cannot as yet tell us exactly what masses all of the 'extra' particles ought to have. Some estimates set the mass of the photino, for example, as low as 250 to 500 eV. That is low enough to make them relativistic, and involve them in the helium production process in the same way that the different flavours of neutrino are involved. If such particles do exist, then they must certainly have participated in the interactions going on in the fireball. But what cosmology and the observed helium abundance reveal is that there can only have been a limited number of such particles (not a limited number of flavours, now, but only a few actual particles) around at the time of nucleosynthesis, compared with the numbers of neutrinos, or they would have influenced the expansion rate of the Universe and led to the production of more helium than we can actually see today. The other relativistic (light) particles are still allowed to exist, but their abundances compared with neutrinos must be very low at the time of nucleosynthesis – 'suppressed', as some cosmologists eloquently put it. And that means that even if these particles still exist in the Universe today, there may not be enough of them to contribute enough mass to push omega right up to one, and close the Universe.

As it happens, one way to 'suppress' the number of such particles is for them to decay into something else, early in the Big Bang. The rate at which such particles decay depends, among other things, on their mass, and in order to make the photinos (or other inos) decay quickly enough they have to be assigned a mass of a few thousand MeV (a few GeV; that is, a few times the proton mass). In that case, of course, they wouldn't contribute in quite the same way to the expansion of the Universe. Particle physicists are quite happy to assign a mass of a few GeV to the photino. Quite apart from this cosmological requirement, some of the particle theorists do favour a mass close to that of the proton, to get the most symmetry in their equations, and the cosmology may well be telling them (and us) which of the options left open by particle theory is the right one to choose. But we don't have to worry too much about the particle physics here. The important point is that the only relativistic light particles which are allowed to fill the Universe are the three flavours of neutrino. Photinos, and other inos, must either be present in very small quantities compared with neutrinos, or much more massive, or both.

The particle physicists, however, still have some tricks up their sleeve. The same theories also predict the existence of heavier particles, inos with masses from a few GeV up to a few thousand times the mass of the proton. Like protons themselves, such heavy particles would not affect the argument about the expansion rate of the Universe and the production of helium, because they are cold – they do not move at relativistic speeds. And there is one oddball particle, called the axion, which particle theorists introduced in the mid-1970s to explain some of the features of the interactions they observe, rather in the way that the neutrino was first introduced as a hypothetical particle, to explain observations of the way neutrons decay. The axion could be very light indeed, perhaps with a mass of only one hundred-thousandth of an electron Volt. But the way in which it is created keeps it 'cold' in the relativistic sense, so that axions do not have the high, relativistic velocities typical of electrons and the light inos (perhaps including photinos) in the early Universe, and do not influence the expansion rate.

There is no shortage of candidates for the dark matter now known to dominate the Universe. But until some experiment here on Earth does detect one of these dark matter candidates, we need not worry too much about exactly which ones are which. What matters is that they come in two principal varieties. First, there are the hot, dark matter particles. The constraint imposed by the helium abundance tells us that the *only* relativistic particles that can be important on a cosmological scale now are the three known flavours of neutrino – although other light inos might be important on smaller scales, perhaps in galaxies or small groups of galaxies. The electrons have long since settled down to a quiet life inside stars and clouds of gas and of dust, where their mass is included in our loose use of the term 'baryonic matter', while the photons, although still relativistic, have no mass to contribute. If one or more of the neutrino flavours had even a small mass, that would certainly be sufficient to make the Universe closed, because there may be as many neutrinos of each flavour as there are photons in the Universe.

Then there are the cold, dark matter candidates, which include the one-off cold, light particle, the axion, and a variety of more massive particles. Happily for the cosmologists, even without knowing exactly which particles are involved, they can distinguish between the effects of hot and cold dark matter in the Universe at large, by looking at the dynamics of galaxies and clusters of galaxies. Once again, as with the limit on the number of possible neutrino flavours, it turns out that cosmology and astrophysics can tell the particle physicists what their experiments here on Earth are likely to find. So the search for the missing mass now takes us back into the realms of 'traditional' astronomy.

The halo conspiracy

Dark matter dominates the Universe. At least 90 per cent of everything, and perhaps 99 per cent, has never been seen. The bright stars and galaxies are not even the tip of the iceberg of the material Universe, since at least all of an iceberg is made of ice, and is attached to the tip. We don't know what the dark matter is, or where it is; only that, unlike stars, it cannot be baryons and that there is enough of it to place the Universe close to the critical dividing line between being open and being closed. The search for the missing mass is now intensifying, with both observers and theorists making their contributions to the debate. But it is already possible to narrow the search down, to eliminate some of the candidates, and to get a good outline idea of what kind of particles it is that really dominate the Universe. Studying galaxies as they are today provides important clues.

First, *where* is the dark matter? The simplest guess would be that it is all in galaxies, after all. Perhaps there is a lot more material, in the form of dust, planets, black holes or something even more exotic, than we see as bright stars. In fact, to bring the Universe close to critical it would mostly have to be something exotic, since all the other possibilities are made of baryons. In recent years, observers have found evidence that there *is* a great deal of dark matter in spiral galaxies – in some cases, at least four times as much dark matter as there is in the form of bright stars. But this is far from being enough to push omega close to 1; indeed, you can just about account for all of this galactic dark matter within the limits allowed for baryons.

These are new discoveries, made only in the past two decades. They depend on measuring the speed with which spiral galaxies rotate. Of course, astronomers cannot watch the changing pattern of stars as a galaxy turns slowly about its nucleus; it takes hundreds of millions of years to complete one turn. But when they look at spiral galaxies that happen to be oriented edge on to us in space, so that they are seen as thin discs, astronomers can use the Doppler shift technique to measure how fast the stars on one side of the disc are moving towards us, and how fast the stars on the other side are moving away, as the disc rotates. Modern spectroscopic techniques are so sensitive that they can measure these Doppler shifts at different distances out from the centre of the tiny image of a galaxy captured by a large telescope, and thereby give velocity measurements across its disc. In recent times these measurements have been extended further out, beyond the region of bright stars, to measure the velocities of clouds of hydrogen gas, still part of the disc of the distant galaxy, using radio astronomy techniques. The results were startling, when they began to come in in the early 1980s.

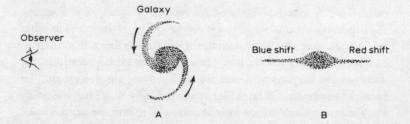

Figure 10.5 When we view a spinning spiral galaxy edge on, we can calculate how rapidly it is rotating by measuring the redshift and blueshift at different distances from the central bulge.

When astronomers plot out the velocities of the stars and clouds orbiting in the disc of a distant galaxy at different distances from its nucleus, they obtain what they call a rotation curve. These curves are usually very symmetrical. The stars on one side of the distant galaxy, at a certain distance out from the centre of that galaxy, are moving towards us at exactly the same speed as stars at the same distance from the centre but on the other side of the nucleus are moving away. That wasn't so surprising. The surprise is that outside the very innermost regions of the galactic centre, on either side and in virtually every case, the speed with which the stars are moving is the same all the way across the disc, as far as measurements can be taken. The rotation curves are very flat, or as some astronomers quip, the most striking feature of rotation curves is that there are no striking features.

This came as a surprise because astronomers had assumed that the greatest amount of mass in a spiral galaxy was concentrated where the galaxy shone brightest, in the central nucleus where there are many stars. If that were the case, then stars further out from the centre should be moving more slowly in their orbits, in exactly the same way that the outer planets of our Solar System (Jupiter, Saturn, Uranus, Neptune and Pluto) move more slowly than the inner planets (Mercury, Venus, Earth and Mars). This is a straightforward conse-quence of the law of gravity discovered by Newton. There are only two ways to account for the flatness of the rotation curves of spiral galaxies. Either Newton's law is wrong (a possibility which has been actively considered by some researchers, but seems rather a drastic assumption) or there must be a great deal of dark matter spread in a huge, roughly spherical halo around each spiral galaxy, and dragging the bright stars round with it as it rotates. Most of the mass, in other words, *cannot* be associated with the bright stars in the central nucleus.

This is the dark matter that can – just – be explained within the allowance of baryons provided for by Big Bang cosmology and the helium abundance. Galaxies like our own Milky Way (and including our Galaxy) are embedded in vast haloes of matter, which might be in the form of objects like large planets ('Jupiters') or failed, dim stars ('brown dwarfs'), or black holes (or it *might* be something more exotic). For some reason, which nobody yet understands, the amount of matter in the halo exactly balances the expected fall-off of rotational

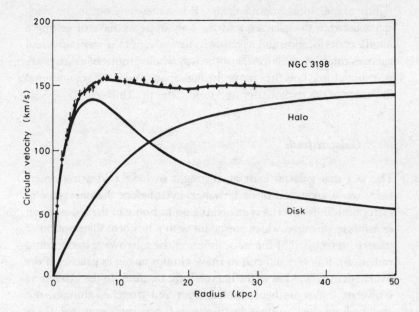

Figure 10.6 The actual rotation curves of galaxies such as NGC 3198 are very flat. This can only be explained if the amount of dark matter in the halo of the galaxy rises, as indicated, while the amount of matter in bright stars falls off outside the central bulge. The two effects 'conspire' to produce the observed rotation curve. (Figure supplied by Tjeerd van Albada.)

velocity from the centre of each galaxy, to produce the featureless rotation curves – a 'conspiracy' that astronomers believe must be more than a coincidence, but which they are at a loss to account for.

Dark matter dominates galaxies, just as it dominates the Universe, and unravelling the way dark matter and bright stars interact in galaxies will keep many astronomers busy for many years. But this is only an incidental feature of the search for the mass needed to close the Universe since, even with all the dark matter required by the rotation curves, galaxies can only account for a small fraction of the mass needed to take the Universe close to critical. These studies help to show that wherever the bulk of the missing mass is, it is *not* tied tightly to individual bright galaxies. It is somewhere out in the black space between the galaxies, and the only hope we have of getting a handle on its location and its nature is to study, not the way individual galaxies rotate on their axes, but the way whole groups of galaxies are distributed, and how they move through space under the gravitational influence of the dark matter that dominates the Universe.

Galactic froth

The fact that galaxies exist at all ought to have told astronomers there was a great deal of dark matter, even before the new wave of astronomical discoveries rammed the point home in the 1980s. In an expanding Universe, which starts out with a uniform distribution of matter (as required by the smoothness of the microwave background radiation), it is very difficult to make clumps as big as galaxies if the overall density is as low as the baryon limit requires. In the expanding Universe, things are being pulled apart and stretched thinner, not clumped together. So how do clumps as big as galaxies grow? There must be some deviations from perfect uniformity early on, some regions where there happens to be a little extra material, and others where there happens to be a little less. Once a region of extra density reaches a certain size it will continue to grow, as its gravity tugs in more matter from outside, holding it back against the general expansion. But if the Universe has only the density implied by the baryon limit, then by the time the original hot gas has cooled to the point where gravity can begin to hold clouds of gas together, it is already so thin that no realistically likely density fluctuation will be big enough to do the job of making a galaxy.

In the 1970s, there were two lines of attack on the problem of how galaxies formed. One view, espoused by Jim Peebles in the United States, held that things had grown from the bottom up. Without knowing how the first 'seeds' formed and grew, he suggested that galaxies formed first, and that only afterwards did galaxies clump together to form clusters of galaxies. Since galaxies are very old – the oldest stars in our Galaxy are almost as old as the Universe itself according to our best models – this idea clearly makes sense. But it makes a prediction, that galaxies and clusters should be distributed randomly throughout the Universe, which is not borne out by modern observations.

The alternative to Peebles' 'bottom up' scenario in the 1970s was the 'top down' idea, purveyed by Yakov Zel'dovich of the Soviet Union. He suggested that in the heat of the early Universe any small-scale fluctuations would have been wiped out, and only very large-scale inhomogeneities would survive as the Universe began to cool out of the Big Bang. On his picture, the original 'seeds' of the structure we see today were not galaxy-sized, but would have been equivalent to superclusters of galaxies, containing a thousand times more mass than the galaxy-sized seeds of Peebles' scenario. When these seeds collapsed into flat pancakes, said Zel'dovich, galaxies would form around the edges of the pancakes, and at regions of extra-high density where two pancakes crossed. He predicted that the Universe would be found to contain strings and chains of galaxies, strung out along filaments like beads on a wire, with great, empty spaces in between the filaments. This is much closer than Peebles' prediction is to the way we do see galaxies distributed today, which gives a boost to the pancake theory. But since galaxies form late in this picture, it is hard to see how there has been time for stars as old as those of our Galaxy to have been born in a Universe as young as the Big Bang calculations imply.

Neither of the two rival theories of the 1970s really works, but they provided a jumping-off point for observers seeking to test the rival ideas, and then for theorists seeking to explain what the observers found using new ideas which go beyond those of the 1970s models.

The most striking feature of the new surveys of the Universe, three-dimensional maps of the distribution of galaxies based on red-shift data, is that the pattern is full of holes. Careful mapping of galaxies with redshifts going up to the equivalent of distances beyond a thousand million light years shows that almost all galaxies congregate around the edges of great bubbles, up to 150 million light years across, which contain little or no luminous matter. Observers see the same picture wherever they choose to carry out their surveys; but unfortunately it is only possible to survey in the necessary detail a very small part of the sky. This has left plenty of scope for speculation about the

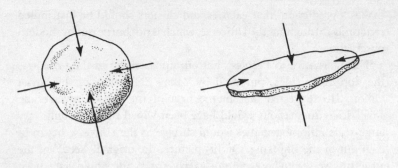

Figure 10.7 Pancake Theory
One view of the way galaxies formed involves the collapse of
huge clouds of material into flat pancakes. Each pancake would
then break up into many galaxies, forming a cluster or supercluster.

exact relationship between the voids and the galaxies, puzzles which
smack of hair-splitting at first sight, but which actually address fundamental questions about the nature of the Big Bang and the location
of the missing mass.

The key question is, do the regions of bright matter, the chains and
shells of galaxies, surround the voids, or do the voids surround the
bright matter? Putting it in slightly more familiar terms, we can think
of the pattern of galaxies on the dark sky as like a pattern of polka
dots; but is it a pattern of white dots (galaxy clusters) on a black
background, or one of black dots (voids) on a white background? In
the mid-1980s, a team of American researchers tackled the problem
by using a computer to study how the bright and dark regions of the
Universe twine about one another – their topology. Richard Gott,
Adrian Melott and Mark Dickinson discovered that *neither* of the
simple guesses is correct. The pattern is not one of black dots on a
white background, nor is it one of white dots on a black background.
Instead, the two structures are completely interconnected, like the
structure of a natural sponge. Strictly speaking, there may be only
one 'void' and one galaxy filament, twined around each other in a
complex fashion.

This topology explains many puzzling features of the observed

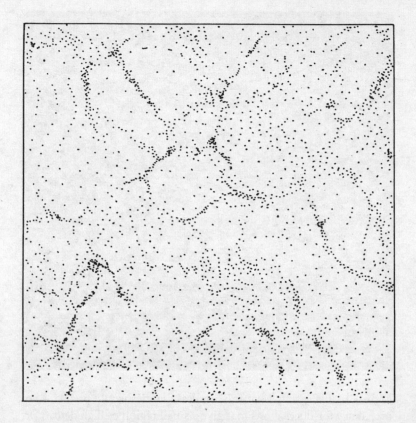

Figure 10.8 Computer simulations of the pancake process show how lines and chains of 'galaxies' would look on the sky, as we viewed the broken-up pancakes edge on.

pattern of galaxies on the sky, which is a two-dimensional projection of a complicated pattern in three dimensions, producing a messy appearance of interconnected clusters and voids mixing into one another. The discovery is very encouraging for the theorists, because it suggests that there is no real distinction between the regions of higher-than-average density where galaxies form and the regions of lower-than-average density, the voids. If the present-day structure of the Universe is simply a result of random fluctuations in the initial

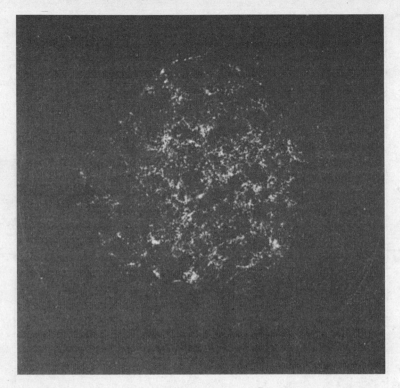

fireball, that is exactly what the simplest theories would predict – no preference for fluctuations that give a small region of high density *or* for fluctuations which produce similar regions of low density, but both kinds of fluctuation occurring at random. The Universe *looks* frothy, but is not, in fact, organized into regular cells, like the structure of a honeycomb.

Melott has taken the idea one step further, and has studied not just the 'snapshot' picture of the distribution of galaxies today, but also the way they are moving in three dimensions. Astronomers can only detect movement of galaxies along the line of sight, the redshift effect; Melott has used computer models to create patterns of simulated 'galaxies' moving in three dimensions, under the influence of the gravity of their neighbours, and has then converted these into the patterns that would be produced for velocities along the line of sight

Figure 10.9 (More than) A Million Galaxies
By combining the information from more than a thousand photographic plates, astronomers have been able to produce these images, showing more than a million galaxies over the entire northern hemisphere sky (left), and more than another million over a large part of the southern sky (right). The overwhelming visual impression is of a network of interconnected chains and filaments of galaxies. How real are the filaments, and what can they tell us about the distribution of matter across the Universe? (From M. Seldner, B. R. Siebers, E. J. Groth and P. J. E. Peebles, *Astronomical Journal*, Volume 82, page 249.)

seen by a hypothetical observer riding on one of the galaxies. He finds that the right kind of pattern, to match observations of real galaxies in our Universe, only comes out if his simulated universes contain enough matter for omega to be close to 1 – but the one-dimensional view from any of the simulated galaxies always gives a false impression that the simulated universes contain less matter than this. It may be that at least some of the missing mass may be accounted for by a kind of cosmic optical illusion, which makes the standard redshift tests give us a false value for omega.

The frothy appearance of the Universe, like so many other factors, is pointing to a density close to critical; it is also very much in line with the idea that galaxies are a result of small inhomogeneities dating back to the birth of the Universe. The next step in the search for the missing mass is to attempt to find out whether it is all associated with the chains and filaments of galaxies that surround the dark voids, or whether it is segregated from the bright matter, and lurks in the voids themselves.

Blowing hot and cold

When astronomers first became convinced of the need to find dark matter to account for the dynamics of galaxies and the Universe, the best candidate seemed to be the sea of neutrinos that they already knew filled space. At least the astronomers knew for sure that neutrinos existed – they are now a completely routine component of particle interactions studied both by theorists and experimenters using the big accelerators. The Big Bang models say that there must be an enormous number of neutrinos in the Universe, ghosting on their way without interacting very much with anything. All it needs is for each of them to have a tiny mass, one billionth of the mass of the proton, and they would provide all the matter needed to close the Universe.

There was a brief burst of excitement, just at the beginning of the 1980s, when two experiments in laboratories on the surface of the Earth came up with hints that neutrinos might indeed have just such a tiny mass. Since then, however, the tide has turned against neutrinos as the explanation for the dark matter. Those first hints of a possible neutrino mass, which require difficult and error-prone measurements, have not been borne out by further tests, and the issue of neutrino mass is still open, taxing experimenters around the world. Even more damningly, though, neutrinos do not, after all, seem to have the right properties to be able to account for the pattern of galaxies seen across the Universe.

In this connection, neutrinos are the archetypal example of the kind of dark matter that cosmologists label 'hot'. Individual neutrinos – and other hot particles – emerge from the Big Bang with very large velocities, close to the speed of light (relativistic velocities), and stream across the Universe freely in all directions. High velocity particles, free-streaming in this manner, tend to smooth out fluctuations in the density of other kinds of matter as the Universe cools – as Jack Burns, of the University of New Mexico, has graphically put it, this is 'much as a cannonball moving at high speed might scatter a loosely built wall of stones without being appreciably slowed by the collision'. In spite of their light mass, such hot neutrinos carry enough energy and momentum, because of their relativistic speed, to break up small-scale structures in the early Universe. This process would continue until the neutrinos cooled to the point where their speeds were, on average, about one-tenth the speed of light. From then on, density fluctuations could grow as the Universe continued to expand, producing a 'top down' pancake universe rather like the earlier ideas of Zel'dovich.

Among other things, this simple hot dark matter scenario implies that there is little or no baryonic matter in the voids between bright galaxy superclusters, and that it has all been swept into the filaments and shells, or bubbles, of the froth. Some astronomers are now trying to probe the dark depths of the nearby voids, using the most sensitive telescopes and detectors to see if there are any galaxies to be found there. Already, there are hints that although the voids are much more sparsely populated than the froth, they do contain some dim galaxies, called dwarfs. That is bad news for all hot dark matter models, including the neutrino-dominated universe. There is also the seemingly insurmountable problem of the ages of stars in our own Galaxy, which don't allow enough time after the Big Bang itself for all the free-streaming and pancake collapse to take place before stars form. In round terms, galaxies must have formed only one or two billion years after the Big Bang, but computer simulations of the pancake process suggest that it needs up to four billion years to do the job. Sadly, even though it is such a straightforward and natural way to account for the large-scale distribution of bright matter across the Universe, astronomers have been forced to accept that at the very least hot dark matter is not the whole story, and that it may be a complete red herring.

The natural alternative to his scenario is to imagine that the Universe is dominated by particles that move much more slowly through space, and therefore do not destroy small-scale fluctuations in density before they have a chance to grow into galaxies. Such scenarios are called 'cold, dark matter' models (CDM for short). The most obvious problem with these models is that nobody has yet detected any particle that could be a candidate for the cold dark matter. But the astronomers were able to specify the properties such a hypothetical particle must have to fit their bill. It (or they) has to be even more reluctant to interact with baryonic matter than neutrinos are, interacting only very weakly indeed, except through gravity. They must be stable, sticking around to dominate the Universe ever since the Big Bang. And, of course, the particles have to have some mass, in order to interact through gravity at all. So they are often referred to as 'weakly interacting massive particles', or WIMPs, and regarded as real candidates for the missing mass even though they have never been detected. The terms WIMP and CDM are synonymous and interchangeable, in this connection.

Although WIMPs have never been detected, cosmologists seeking the missing mass were not entirely guilty of pulling a rabbit out of a hat when they suggested the Universe might be dominated by cold dark matter instead of neutrinos. Particle physicists have been developing a very successful unified theory of how the forces and particles of nature interact. This theory, called supersymmetry (SUSY) accounts for what we know about the particle world, but at the cost of requiring that for every kind of particle we already know about there must be a supersymmetric partner. For example, the electron has a (hypothetical) counterpart dubbed the selectron; the photon a partner called the photino; and so on. Most of these particles are of no concern to the search for dark matter, since they are unstable and will rapidly decay into other particles (assuming they really do exist at all!). But the theory requires that there should be one type of SUSY particle (the lightest member of the family) that *is* stable, and stays around forever, and probably one that is very long lived. They would have very small masses, but it is a clear requirement of the basic theory that there *is* dark matter in the Universe, and that its overall contribution to the density cannot be ignored. As I have mentioned, there is also scope for a slightly different kind of particle, called the axion, whose presence would help the particle physicists to explain some subtleties of the interactions they observe, but which, like the SUSY particles, has never been directly detected. Armed with all that information, many cosmologists have been happy to assume that WIMPs exist in sufficient numbers to close the Universe, even though no SUSY particle has yet been detected.

Assuming WIMPs do exist, what would a CDM-dominated Universe look like? The models have no difficulty accounting for the existence of features like galaxies and clusters of galaxies. As the slow-moving WIMPs congregate in clumps in the expanding Universe, the clumps will exert a strong gravitational pull on any baryonic material nearby, acting like deep holes into which the baryons fall. The first clumps are smaller than in the hot dark matter scenario, so that galaxies are older than clusters, which are in turn older than superclusters. Which is where the model runs into difficulties. In the hot model, there doesn't seem to have been time for galaxies to form, but the large-scale pattern of bright matter across the Universe falls out fairly well from the equations. In the simple cold model, it is easy to form galaxies, but there hasn't been time for clusters of clusters of galaxies to group together in long chains and filaments. Perhaps the truth lies somewhere in between – although even the imaginations of the particle physicists haven't yet found a way to produce 'warm' dark matter out of the Big Bang – or in a combination of two or more different kinds of dark matter. Or, perhaps, there is a different solution to the puzzle, a variation on the basic CDM theme.

When cosmologists map out the patterns of high and low density across the 'Universe' in their computer models, they are really tracing the distribution of the dark matter, which is 99 per cent of all the gravitationally important material around. Perhaps the dark matter is distributed more smoothly than the clumpy distribution of bright matter, with lots of unseen material in the voids between the chains and filaments of the froth. It may be that baryons falling into gravitational potholes only light up to form galaxies of stars when they fall in to the very deepest potholes, the regions where the dark matter is concentrated most strongly, and that in most of the Universe there are huge clouds of hydrogen gas gripped in the gravitational embrace of the WIMPs, but not strongly enough to trigger the processes that make galaxies.[8] It would help if we knew exactly what those processes were, but we do not. Even without that information, though, it is at least worth considering that by concentrating our attention on bright galaxies, we may be getting a completely biased picture of the overall distribution of matter through space.

Until WIMPs are detected here on Earth, or (much harder) proved not to exist, or until completely new observational evidence comes in, many of these questions will remain open. But it seems to me that the cold, dark matter scenario, in one form or another, is by far the best model we have. It is certainly imperfect, and will be modified considerably in the years ahead. It is possible that it will turn out to be completely wrong, but that now seems very unlikely. And since space does not allow me to go into details of all the models of dark matter universes now being discussed by the theorists, it seems best to concentrate on the front runner, with the warning that the race is not yet over, that the front runner may fall, and that although my personal prejudice is in favour of the WIMP model, I have been known to back astronomical losers in the past!

A bias for cold, dark matter

The way astronomers test their ideas about galaxy formation in models of the Universe dominated by different kinds of dark matter is with computer simulations. We can see the pattern of galaxies on the sky, and the kind of chains and filaments the pattern produces. In the computer models, equations are set up to describe points representing galaxies, themselves set up in a uniform three-dimensional cubic grid. The computer program then does the numerical equivalent of displacing the 'galaxies' slightly from their starting positions and allowing them to interact in accordance with the law of gravity, while the grid is expanded to simulate the expansion of the Universe. The interactions between the 'galaxies' can be altered by putting in the effects that would be produced by hot, cold or warm dark matter, and after a suitable interval of computer time, representing the evolution of the Universe up to the present day, the new positions of the galaxies, now distributed far from uniformly, can be shown on a screen, or printed out, as representations of how the sky would look in such a model Universe. At a superficial level, the patterns that are most like those of the real sky can be picked out by eye; more subtle tests, involving precise statistical comparisons of the nature of the chains and filaments that are produced by the computer simulations with the chains and filaments of the real world, give the ultimate test of whether each model is a good or bad approximation to reality.

All this takes up a great deal of computer time. There are several groups involved in such work; I discussed the basic techniques with Carlos Frenck, a Mexican astronomer now at the University of Durham, in England, who has carried out such studies in collaboration with George Efstathiou, Marc Davis and Simon White. One of their studies involves a grid of 32,768 'galaxies', perturbed by the appropriate interactions. This may seem like a modest representation of the real Universe, where over a million galaxies can easily be pictured by putting together images from different plates of the northern sky alone (see Figure 10.9). But it does give a powerful insight into how the Universe works.

These 'N-body' simulations (in this example, 'N' is 32,768) mesh in neatly with the calculations made by the theorists of how matter ought to behave in a Universe dominated by WIMPs. Objects with masses in the range from a hundred million times the mass of the Sun to a trillion (a million million) times the mass of the Sun would have formed quickly in a CDM Universe, and this is just the range of masses occupied by known galaxies, which are indeed just about as old as the Universe itself.

The galaxies are supposed to have formed from local density fluctuations in the sea of WIMPs, producing gravitational potholes which trapped the baryonic material. Most galaxies are spirals, like our own, and presumably correspond to modest, but relatively common, fluctuations. Giant, elliptical galaxies are rare, certainly no more than

15 per cent of the total, and would have formed from larger, but rarer, random fluctuations in primordial density. This is not just a rule of thumb. There are precise mathematical equations describing the nature of such random fluctuations, and the exact proportions of large and small galaxies turn out to follow those statistical rules very well indeed.

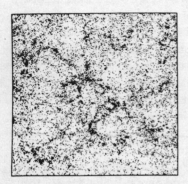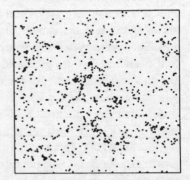

Figure 10.10 When the computer simulations are carried out using cold dark matter and with omega equal to one, the picture begins to look very much like the real Universe. The overall distribution of matter, including the CDM, in such a Universe is shown on the left; the distribution of 'galaxies', represented as regions where the overall density reaches peak levels, is shown on the right. (These are the 'N-body' simulations described in the text; figures supplied by Carlos Frenck.)

The theory even predicts that as baryonic material falls into a pothole and begins to form stars the gravitational interaction between the baryonic material and the dark matter of what is now a galactic halo will produce an orbital velocity in the developing galaxy that is the same right across its disc, exactly as observed.

Now the observers come into the picture. The cosmic background radiation is very uniform, homogeneous to within one part in twenty-five thousand. It last interacted with matter 500,000 years after the Big Bang itself, so we know that matter was equally uniformly distributed at that time. In that case, the voids *cannot* be empty, because even ten or twenty billion years is insufficient time to sweep matter out of them so drastically as to produce a Universe as inhomogeneous as ours *seems* to be, judging by the distribution of bright galaxies. Mass does not follow the light, and somehow galaxies must have formed only where the density fluctuations were at their biggest, leaving many failed galaxies scattered across the voids. Most matter is distributed in a vast unseen ocean, much more evenly than the visible galaxies; even a huge supercluster of galaxies is only the equivalent of a small ripple on the sea of dark matter.

The computer simulations can reproduce all of the observed features of the bright galaxy distribution beautifully, but only for either of two possible scenarios, each involving cold, not hot, dark matter. In the first, the model Universe is open, with $\Omega = 0.2$ and the galaxies assumed to be a fair tracer of the mass. In the second, $\Omega = 1$, the Hubble parameter has a value of 50 km/sec/Mpc, and the galaxies must be more strongly clumped than the dark matter. Quite apart from any prejudices in favour of $\Omega = 1$ in the light of the evidence discussed earlier in this book, the microwave background evidence seems to tilt the balance strongly in favour of the second option. As icing on the cake, when Frenck and his colleagues looked at galaxies themselves they found more evidence that they were on the right track. Once the details of the model are fixed by making it match up with the observed Universe, these details can be used as parameters which limit the behaviour of galaxies within that model, in a new set of computer simulations. The simulation with $\Omega = 1$ and biasing automatically produces the massive haloes and flat rotation curves that were such a surprise to astronomers just a few years ago, with no further effort by the modellers. Frenck stresses that this is 'quite remarkable' since the 'free parameters of the model had been fixed beforehand by observations'. All that's left, it seems, is to explain why galaxies formed in this biased way, and failed to form in the voids.

Astronomers only recently realized the need for such biasing to be operating, so relatively little work has been done on what the processes involved might be. But as soon as they began to devote a little thought to the problem, theorists realized, once again, that the observations were rubbing their noses in something that ought to have been obvious all along. When the first galaxies formed, in the biggest gravitational potholes, they may well have exploded into life as stars lit up one after another. This process of primordial star formation could have sent a blast of energy outward from each young galaxy, energy carried both in the form of electromagnetic radiation (heat, light, ultraviolet radiation, X-rays and so on) and as a blast wave through the material of intergalactic space. Both ultraviolet radiation and a 'wind' of energetic particles ejected from forming galaxies turn out to be quite efficient at suppressing the formation of other galaxies over a range of 20 Megaparsecs or more. The trick works by providing more energy for each of the baryons in the intergalactic medium – by making the material hotter – so that it does not settle into the potholes so easily. All it needs is for less than 10 per cent of the total baryon mass to be converted into galaxies early on, in the deepest potholes, with less than 1 per cent of the energy produced as a result spreading out to the baryons in the voids.

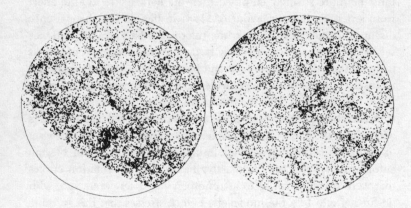

Figure 10.11 A direct comparison of the real distribution of galaxies on the sky (left) with a computer simulation for $\Omega =$ 1 in a cold, dark matter Universe. (Figures supplied by Carlos Frenck.)

Martin Rees, of the University of Cambridge, has stressed that these requirements are easily met, and that the proposed mechanisms are not weird inventions of fevered theorists trying desperately to shore up a shaky model. It would be astonishing, he says, if none of the processes now being investigated were important – that is, if there were *no* large-scale environmental effects that influenced galaxy formation. In spite of the uncertainties at present, there is certainly no need for proponents of the idea that omega is equal to 1 to be embarrassed by the frothy distribution of bright galaxies across the Universe, says Rees.

But I should mention some of the other ideas, far stranger than WIMPs, which are waiting in the wings to come forward and provide the missing mass if WIMPs do prove inadequate to the task – or, perhaps, to help the WIMP models out of their difficulties by combining with the cold dark matter to explain details of the galaxy distribution. Just maybe one, or more, of these ideas will turn out to have a role to play in the real Universe.

Strings and things

The most respectable of the wild ideas concerns the possible existence of cosmic string. But unfortunately, that terminology has been used in three different, but possibly related, contexts, so it's best to unravel the three uses of the term before getting on to the cosmological nitty gritty.

Working from the smallest scale upwards, many theorists are excited today about a relatively new theory of the particle world. Most of us have a mental image of particles like quarks and leptons as little round balls. For years, the physicists have been telling us this is wrong, and that we should think of these fundamental entities as mathematical points, occupying no space at all in any direction, but extending their spheres of influence over the range of the fields of force associated with them.[9] Now, though, theorists seeking to unify all of the forces of nature into one mathematical description have found that they can overcome many of the difficulties by working in a 10-dimensional spacetime in which the fundamental entities are not represented by mathematical points but by tiny objects which have a one-dimensional length, little loops extending over about 10^{-35} (a decimal point followed by 35 zeroes and a 1) of a centimetre. The 'extra' dimensions, over and above our familiar three of space and one of time, are wrapped up in these tiny tubes. With grand hyperbole, the theory that builds from this, the smallest scale yet considered by scientists, is called 'superstrings'.

Superstring theory includes within itself the possibility of another kind of string, existing on a cosmic scale. But these cosmic strings, which are especially interesting in the context of the debate about dark matter, are not exclusively a product of superstring theory. Many other variations on the theoretical theme of the behaviour of matter and energy at the time of the inflationary era of the birth of the Universe produce the same prediction – that there ought to be almost infinitesimally thin threads of mass-energy stretching across the Universe, each one about 10^{-30} centimetre across but enormously massive, containing about 10^{20} kilograms (a hundred thousand trillion tonnes) of matter in every centimetre of string. Such strings are under equally enormous tension, and if they are stretched then the energy used to stretch them is converted into mass, so that they become even more massive.

The way in which such exotic objects might be produced depends on the way the original unity of the Universe was broken up at the earliest times. The Grand Unified Theories (GUTs) of physics suggest that under conditions of very high energy there is no distinction between the fundamental forces of nature. Electromagnetism and the forces that dominate inside atomic nuclei, the strong and weak forces, are today very different from one another, and all three are different from gravity. But theorists argue that at the moment of creation itself all four forces were equally strong and governed by one set of mathematical rules. As the Universe cooled, the different forces separated out and showed four different faces to the world, but those four faces should still be describable in terms of one unified mathematical package. The fact that theorists have not yet found the right mathematical description for that unified package is regarded as a minor inconvenience; they have several good lines of attack on the problem, including superstring theory, and they are sure they will crack it one day. The important point, as far as we are concerned now, is that in many variations on the grand unified theme the splitting off of the four forces as the Universe cools involves a change from one state to another, a fundamental alteration in the physical properties of the Universe, which is called a phase transition. I mentioned earlier (page 229) that this is likened by physicists to the phase change that occurs when water freezes into ice. Now, it is time to go into a little more detail.

When a liquid crystallizes, changing into a solid, it often does so imperfectly. Instead of one uniform lump of solid crystal, there may be separate regions, separate domains within the crystal, which are perfectly smooth and uniform within themselves, but in which the atoms and molecules are aligned differently from the pattern in the domain next door. So there are boundaries between domains, faults in the crystal, sometimes called discontinuities. Cosmic strings are faults in the fabric of spacetime, caused by imperfections in the phase transition that cooled the Universe out of its grand unified state. In a sense, these narrow but enormously long tubes enclose regions of spacetime in which the rules of grand unification still apply, and where all the forces of nature are one. They were produced, according to the grand unified theories, a mere 10^{-35} second after the moment of creation itself.

Like WIMPs, cosmic strings should have been produced in vast numbers in the Big Bang. Unlike WIMPs, they whip through the Universe like highly stretched rubber bands, twanging at close to the speed of light. But most of them would not have survived to the present day. Such strings have no free ends, but always loop back on themselves to keep the string closed off from the rest of the Universe; the loops, however, can stretch across the entire visible Universe. But when two pieces of cosmic string cross (either two separate pieces or a tangled loop from one long piece of string), they can make new connections and shed smaller loops. The vibrating smaller loops also radiate energy away by gravitational radiation, making the loops shrink. Small loops are cut off from large loops which are cut off from huge loops which are . . . but you get the picture. If cosmic strings exist, then the Universe contains loops of material with various masses, just right to act as seeds for the growth of galaxies through their gravitational influence on the surrounding sea of WIMPs and baryons. Small wonder that astronomers seeking to account for the presence of the third kind of universal string, the long chains of galaxies that I have been careful to refer to so far as filaments, have seized with delight on the idea of cosmic string. Could it be that superstring begat cosmic string which begat strings of galaxies? Even with all the

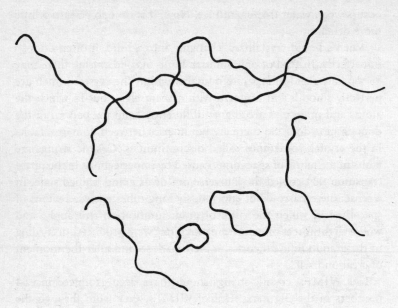

Figure 10.12 Cosmic String
If two cosmic strings overlap, they can break and reconnect to
form loops. Could these be the seeds of galaxies?

uncertainties in the chain of argument, the possibility is intriguing
enough to follow up.

The properties of cosmic string certainly seem to be just what the
astronomers wanted. It is rather strange stuff, because in spite of its
enormous mass a long, straight cosmic string actually has only a small
gravitational influence on its immediate surroundings today. In effect,
though, a straight string takes a slice out of spacetime as it passes,
altering the local geometry with bizarre implications for any object it
happens to meet. Suppose a string passed right through your body,
horizontally at waist height. You wouldn't feel a thing, at first. But in
the wake of the string, where spacetime is distorted, the top and
bottom of your body would be moving together at a speed of about
4 km per second, with uncomfortable consequences. Or imagine such
a string passing close by, through your room but not through yourself.
Again, you would feel no gravitational pull at all, and if your eyes

were shut you wouldn't know anything had happened – until the wall on the opposite side of the room smacked you in the face at a speed of several kilometres a second! Strings moving through space leave a pronounced wake behind them, and it is very tempting to imagine that these wakes might account for the growth of galaxies in filaments and sheets across the Universe. Alexander Vilenkin, of Tufts University, said as much to a meeting debating the links between particle physics and cosmology, at Fermilab in Chicago in 1984: 'A distinctive feature of the string scenario is the formation of planar wakes behind relativistically moving strings. These wakes can help to explain the formation of large-scale structure in a Universe dominated by weakly interacting cold particles, such as axions.'

But there is a limit to how much cosmic string there can be around today. Anything that distorts spacetime bends light around itself, and distorts other forms of electromagnetic wave, such as the cosmic background radiation, passing by. Some astronomers hope to identify cosmic strings far away across the Universe by finding the influence – a kind of long, thin distorting lens effect – in the images of even more distant objects, such as quasars. They haven't yet succeeded, and perhaps there simply are not enough long pieces of cosmic string left for the effect to be noticeable. We know that the Universe is very uniform, from the microwave background evidence. That sets limits on the amount of cosmic string there is, since too much would produce bigger variations in the background from place to place on the sky than we actually see. Present observations of the background tell us that strings cannot contribute more than one hundred-thousandth of the total density needed to close the Universe. But even a fraction one-tenth smaller than this, with strings providing only one millionth of the closure density, would still be enough for them to leave their mark by producing galaxies in a Universe dominated by cold dark matter.

The details are still being worked out, and new ideas are aired almost every month. But it does seem that loops of string might help galaxies to grow. The smallest loops around today would weigh in at about the mass of a billion Suns, and encompass a diameter of about 10,000 light years; such small loops affect spacetime rather differently from a long, straight string, and would indeed tug in matter, both WIMPs and baryons, like water running into a lake. Even better, according to calculations made by Neil Turok, of Imperial College, in London, the way large loops split off smaller loops, and the smaller loops then cluster together, fits very neatly the pattern of characteristic clusters of galaxies, named 'Abell clusters' after their discoverer, which typically contain fifty or more galaxies grouped in a volume of space only some 1.5 Megaparsecs (just less than five light years) across. Even before anyone has yet carried through the necessary calculations to establish the case, it is natural to speculate that bright galaxies may form only in frothy sheets and filaments across a Universe full of cold dark matter because those frothy sheets and filaments happen to be the places where cosmic strings have passed by, and where loops of string remain today at the hearts of galaxies and galaxy clusters.

Cosmic string can even revive the hot dark matter scenario. Although small-scale variations in the density of baryons are still wiped out by free-streaming neutrinos (or other HDM particles) even if the Universe contains cosmic strings, as soon as the neutrinos cool to the point where they are unable to maintain the smoothness both baryons and neutrinos can begin to accumulate around loops of string, forming galaxies much earlier than in an HDM model without string. Division of string loops could, perhaps, explain why some galaxies seem to be in pairs that have split apart, like amoebas; although nobody has yet carried the necessary calculations through, there is speculation that the disturbances produced in the wake of a passing string could account for the large-scale 'streaming' velocities of galaxies, like those recently found in our vicinity; and the now obvious connection of many galaxies in chains and filaments – the third kind of string – could also be explained by the formation of a line of 'seeds' as an original long, tangled string doubles over itself and splits off loops. It will be especially interesting to see how the N-body simulations come out once the effects of string are included in the computer programs; meanwhile, it is hardly surprising that astronomers are excited about cosmic string, which is certainly the current flavour of the decade. A combination of strings *and* WIMPs (or even neutrinos with mass) reproduces so many features of the observed Universe that it is tempting for theorists to believe that they are on the right track at last. And they still have other cards up their sleeve.

Black holes, quark nuggets and shadow matter

Apart from changing the rules of the game, by saying we don't understand gravity after all, or invoking the presence of something we don't know anything about (both desperate councils of despair), there are three more cards the theorists have ready to play if needed. One sounds familiar, but appears in an unfamiliar guise – the concept of black holes.

Black holes would certainly be good candidates for dark matter, and do indeed influence their surroundings through gravity, as required. But 'ordinary' black holes, made when stars die and collapse in upon themselves, or when matter funnels on to a supermassive core at the heart of a galaxy or quasar, cannot be invoked to provide the missing mass, since they are originally composed of baryons themselves. Even if they are later crushed out of existence inside black holes, the baryons were still manufactured in the Big Bang, and are still subject to the limits set by the helium abundance. The only way that black holes could provide the dark matter needed to close the Universe, without running into this limit on baryon numbers, is if the black holes formed *before* the baryons were cooked, even earlier in the Big Bang. Such primordial black holes would be much smaller than atoms, and would each have about the mass of a planet. They *could* form from density fluctuations in the very early Universe, before the epoch of baryons, but the only reason for invoking their existence is to provide the missing mass. There is no observational evidence that they exist, and they do not naturally produce the required frothy distribution of galaxies in the real Universe. As a proposal for the dark matter that dominates the Universe, they scarcely rank above invoking magic, or mysterious unknown phenomena.

If you have a taste for scientific ideas that smack of magic, though, the theorists can provide a much more entertaining possibility than black holes. One variation on the superstring theme, which includes gravity and all the other forces in one mathematical formalism, contains an extra kind of splitting over and above the symmetry breaking that is required to divide the original grand unified force into four components. According to this version of the equations, there was another splitting, even earlier in the Big Bang, which produced two separate sets of particles and forces, each occupying the same Universe. One set is our familiar world of stars, galaxies, planets and the rest, held together by the familiar four forces. The other set is – something else, invisible and perhaps undetectable, co-existing with us in our Universe but with its own particles and forces, interacting only with each other, not with our world.

What would this shadow world be like? It is possible, though unlikely, that it could be a kind of duplicate of our Universe, with the same, or similar, four forces, particles equivalent to quarks and leptons, and shadow stars, planets and even people going about their business and (perhaps) speculating about the possibility that *our* world exists. It is far more likely, however, that *if* the shadow world exists then the laws of physics will be slightly different, or very different, there. It could contain different particles that obey different rules. But either way there is only one way in which the two worlds could interfere with one another, and that is through gravity. Could the missing mass be in the form of a whole alternate world, occupying the same space as our own? Could some of the dark matter of the Galaxy's disc be in the form of shadow stars and planets? Or might our whole Universe be filled with exotic shadow particles that contribute to the overall density of the Universe but play no part in the particle interactions that involve neutrinos, WIMPs, cosmic strings and the rest of our world?

This is far from being the end of the imaginings of the theorists. But shadow matter represents in many ways the most extreme possibility yet conjured up. The notion should be served up with a large pinch of salt, and the better we can explain the observed distribution of galaxies without invoking shadow matter, or other exotic possibilities, the less likely it is that this bizarre vision will turn out to have any practical relevance. Even so, it is comforting to be reminded that if those other ideas, that look so strong today, do turn out to be flawed, there are still theorists around with imaginations vivid enough to dream up new possibilities.

Somewhere on the plausibility scale in between the wild idea of shadow matter and the relative respectability of WIMP is another theoretical speculation, the concept of strange matter. From one point of view, this is a rather conservative idea, since it does not require the presence of any completely new particles, but only the kind of matter we already know about in a new, denser form. The idea which some theorists have followed up is that a lump of matter made up of roughly equal numbers of up, down and strange quarks might be stable. Familiar baryonic matter, remember, contains only up and down quarks; the name given to this hypothetical form of matter containing strange quarks as well is, logically enough, 'strange matter'. Hypothetical lumps of strange matter are sometimes called quark nuggets; there is no certainty that they would be stable – that depends on details of the strong interaction and other properties of quarks that are hard to calculate accurately. But if they do exist, quark nuggets contain more or less conventional matter which has never been processed into baryons as such, and so is not subject to the helium limit. Speculations about strange matter range from the possibility that nuggets with mass less than twice the mass of a proton might have been left over from the Big Bang, and provide the missing mass, to the idea that whole stars of strange matter might exist today, produced by the collapse of old, dead stars through a neutron star state and on into a strange matter state. Even if strange matter exists, however, there is no natural way in which it could account for all of the dark matter required to close the Universe.

The best candidates for the missing mass are not the ones dreamed up by the theorists, but the ones forced upon us by observations and experiments. WIMPs are *required* by the N-body simulations, for example, in order to provide a match with the real world; and cold dark matter is also the best candidate to explain other puzzling astronomical observations. The rest, interesting though it may be to fans of science fiction, is largely conjecture. So perhaps, having wandered so far off the beaten track, I should outline the current cosmological 'best buy' before taking a final look at the broad picture and assessing the ultimate fate of the Universe.

The best buy

The current best buy in cosmology is a combination of the old standard model of the Big Bang, the newer idea of inflation, and the presence of enough cold, dark matter to make the Universe flat. In spite of any premature obituaries you may have read, the most soundly based component of that trinity is still the standard Big Bang; debate in the 1990s about the plausibility of certain details in specific variations on the Big Bang theme does nothing to undermine the overwhelming weight of evidence, discussed earlier in this book, that we live in a Universe that has expanded out of a hot, dense state.

The usual way to make this point is to stress the successful matches between theory and observation – the spectrum of the cosmic background radiation, the abundances of light elements such as helium and deuterium, and so on. But Martin Rees points out that there is an equally forceful example in favour of the hot Big Bang – the fact that no discoveries have been made which invalidate it. In principle, it would have been easy to prove the Big Bang theory wrong. If astronomers had found any stars with zero helium abundance, or objects much older than the inferred age of the Universe, or if it had turned out that neutrinos had a mass of around 1 keV (which would have made the Big Bang model collapse before we ever existed), then the standard model would have been completely ruled out. 'The evidence that everything emerged from a hot dense state, which we can trace back to the time when it was about one second old, should,' Rees said in 1991, 'be taken as seriously as, for instance, ideas about the early history of our Earth.'

But this doesn't mean we know everything – in particular, in spite of the success of CDM simulations, we still don't know exactly how galaxies formed. Extending his geophysical analogy, Rees says that although we know for sure that the Earth is round, we don't know with the same certainty just how the features on its surface, the continents, mountain ranges, hills and valleys formed. Some features seem to be broadly explained by good theories, but some topographic details will never be understood as more than accidents of nature. But the fact that we may not understand how a particular mountain formed does not mean that we have to abandon the idea that the Earth is round. In the same way, the fact that we may not understand precisely how our Galaxy formed does not mean that we have to abandon the Big Bang theory. Even if the entire CDM package were ruled out by some new observations, this would be a disappointment, removing a particularly simple and neat explanation of galaxy formation, but not a blow to the standard model of the Big Bang itself.

But is the CDM model really in the difficulties that some commentators suggested in the early 1990s? Look at the picture again. Our Universe contains matter, some of which we can see in the form of bright galaxies of stars, but most of which is dark and cannot be directly detected. Galaxies are grouped together in clumps, much bigger than astronomers had suspected until recently. An individual supercluster can extend across 500 million light years, and some filaments are so long that a few astronomers believe they may be just the visible portions of strings of galaxies that stretch across the entire Universe. Galaxies within superclusters move much more rapidly compared with one another than anyone had expected, and large numbers of galaxies move together with velocities of hundreds of kilometres per second relative to the cosmic background radiation, quite separately from their motion due to the expansion of the Universe.

The 'best buy' models of the Universe can explain all this. We do not know for sure what 99 per cent of the cosmic mass consists of, and it is likely that it is not all the same stuff. 'Ordinary', baryonic dark matter, for instance, is quite sufficient to explain the way stars concentrate in the plane of our own Galaxy, with brown dwarfs and Jupiters doing their gravitational bit to keep things held in place. But when the growth of structure in the Universe is traced by N-body calculations in the biggest computers available, it turns out that while neutrinos might just be made to fit the bill (perhaps with the aid of cosmic string), most of the observed features of the galaxy distribution fall naturally into place if the Universe is just closed (with omega equal to 1) and the hidden mass is in the form of weakly interacting cold dark matter. The galaxy formation process has to be 'biased' to account for the observed frothy appearance, but there are several likely mechanisms to produce biasing, even without invoking cosmic strings.

One example shows both the public interest in these new cosmological discoveries and the danger of trying to read too much into each new piece of evidence. At the end of 1990, Carlos Frenck and his colleagues published a paper in *Nature* (Volume 349, p. 32) in which they pointed out that the exact pattern of galaxies found in their latest simulations did not quite match the pattern of galaxies found in the IRAS survey. The paper appeared at Christmas, a quiet time for news, and was accompanied by a *Nature* press release suggesting that the arch priests of the CDM theory were abandoning their baby. The resulting wave of publicity for the notion that the CDM theory was dead reached such a pitch that Frenck and Nick Kaiser had to publish a letter in *Nature* (Volume 351, p. 22) pointing out that this was not, in fact, the gist of their original paper.

In that letter, they summed up the standard CDM model as having four components:

First: the dark matter consists of weakly interacting cold particles
Second: the Universe has the critical density
Third: structures such as galaxies and clusters grew on seed fluctuations of the type predicted by inflationary theory
Fourth: 'that the distribution of galaxies is related to the distribution of mass through a simple statistical prescription, the "linear biasing model" '

Hold on to that last one, even though it sounds complicated at first, because that is the one that all the fuss was about. Items one to three, which form the heart of CDM theory, were *not* brought into doubt by the new work. So, what is this linear biasing, and what seems to be wrong with it?

The key number, called the 'biasing factor' is defined in such a way that it is 1 if the distribution of bright galaxies is the same as the distribution of dark matter, and greater than 1 if (as CDM models require) galaxies are more clustered than the dark matter. As you might guess, in hot dark matter models biasing works the other way, and this biasing factor, or parameter, must be much less than 1; in fact, studies of the cosmic microwave background show that the biasing factor is definitely greater than 0.86, which effectively rules out the HDM models and is getting close to the value predicted by CDM models, even though the background measurements cannot yet give a precise value for the biasing factor. The debate is now about just how much bigger than 1 the parameter is, and what you have to do to tweak up the CDM models to make them match the observations.

According to the IRAS data, omega is indeed 1, and the biasing parameter has a value of 1.3. This requires a little more biasing than in the simplest version of the CDM model – but it turns out that the extra biasing required is exactly what should be expected if cosmic strings are playing a part in the formation of galaxies and clusters of galaxies.

There may, of course, be some other reason for the slightly high value of the biasing parameter. But when the problem can be resolved simply by invoking the most plausible of the new ideas already on the table, the theory hardly seems to be in terminal difficulties. Cosmic strings fit neatly into the CDM scenario, and there are no signs of despair among the CDM theorists yet.

But all of this debate is to some extent a diversion from the main theme of this book. It doesn't really matter what the dark matter is, or how it is distributed across the Universe, as far as the ultimate fate of the Universe is concerned. All that matters is which side of the critical dividing line between being open and closed the Universe sits. Even inflation cannot actually tell us this. The Universe might be just open, or it might be just closed. We still have a choice of theoretical futures to consider. But the very latest ideas about the origin and ultimate fate of the Universe suggest that we may be able to have our theoretical cake and eat it, with finite bubble universes like our own forming temporary features within an eternal spacetime foam.

Postscript

At the end of 1997, the current 'best buy' has indeed been tweaked, thanks to the latest computer simulations. Cosmic string now looks less plausible, but the right pattern of galaxies on the sky is produced by models with omega close to 1, made up of about 70 per cent CDM, 30 per cent HDM, and a few per cent baryonic matter. Watch this space!

Chapter 11

The Birth and Death of
the Universe

The ultimate fate of the Universe is already sealed by the initial conditions in which it was born. Eternal expansion, or eventual recollapse, depend on which side of the critical dividing line $\Omega = 1$ the Universe sits, and whichever side it sits it has sat there since time began. But either way, we know that the Universe sits very close to that critical value, so that the future that lies ahead is much longer than the 15 billion years or so of history that lies behind us. For some purposes, it scarcely matters which kind of future is in store. Our own Sun, for example, will swell up to become a red giant star, burning the Earth to a cinder, in about 5 billion years from now, and that is probably as far ahead as anyone interested only in the future of the human race need bother to look.[1] But even this represents merely a hiccup on the cosmic scale of things. Given long enough, whether the Universe is open or closed, matter itself will begin to die.

The fates of stars

For the moment, let's set aside the new discoveries that tell us that 90 per cent of the Universe is not composed of familiar baryons in the form of stars and planets. What will happen to the kind of matter we do know and love as the Universe ages? Stars themselves, though they may burn nuclear fuel to hold themselves up against the pull of gravity for billions of years, cannot last forever. When their fuel is exhausted, gravity must win its long, drawn out battle, and cause the stellar remnants to collapse. And, it turns out, there are only three things that a dead star can collapse into.

Stars form from clouds of dust and gas in space, by processes which are still not very well understood by physicists. But it is very clear how they get hot, and stay hot, once they have formed. Gravitational collapse releases heat which makes the young star glow brightly, and as the star collapses and becomes more compact and dense that heat is enough to initiate fusion reactions in its heart, like the ones which keep the Sun hot today. The more massive a star is, the more fuel it must burn each second to hold itself up against the pull of gravity. Some stars exist in this stable main sequence state for only a few million years; the Sun has already done so for 4.5 billion years, and has about as long still to go before its hydrogen fuel is exhausted; other stars may live for even longer. Things must change, though, when all the hydrogen in the core of a star has been converted into helium.

With no more hydrogen to burn, the star is no longer able to resist the pull of gravity, and the core begins to collapse once more. But this releases more gravitational energy as heat, and new nuclear reactions begin as a result, converting helium into carbon. The energy released in the process makes the outer layers of the star expand, and this is why the Earth will be crisped by the Sun in about 5 billion years from now. Eventually, the helium will all be converted into carbon, and the process will repeat as before, with the core getting more compact and hotter still, while carbon is converted into oxygen by fusion. The process can repeat several times, until most of the core material has been converted into iron-56. But there it must end, because the addition of more protons and neutrons to the nucleus of iron-56 does not release energy. Instead, in order to build up heavier elements energy must be put *in* from somewhere.

In some large stars, that is exactly what happens next. With its source of heat at last removed, the star collapses, with the mass of the outer layers – its distended atmosphere – falling inwards under the pull of gravity, and themselves getting very hot as gravitational energy is released. In these conditions, given enough mass, there can be a sudden explosion of fusion activity as hydrogen and helium from the atmosphere of the star squeeze down on to the core. The explosion takes place in a shell around the core, like the peel surrounding an orange, and the resulting blast travels in two directions – outward, ejecting the rest of the atmosphere of the star to form a glowing, expanding nebula, and inward, squeezing the core tight and also producing a smattering of elements heavier than iron, some of which may get thrown out into the new nebula. The star has become a nova, or a supernova. In its death throes, it has helped to seed the interstellar medium with heavy elements that will go into the next generation of stars and planets. It is only because previous generations of stars have gone through this cycle that our Sun has a family of planets at all, and we are here, with our bodies based on carbon chemistry and breathing oxygen out of the air, to puzzle over the origins of clouds in space like the famous Crab Nebula.

Not all stars, however, end their active lives in such spectacular fashion. Less massive stars, like our Sun, may fade away more quietly, puffing out a little gas into space but then settling down gently as glowing cinders once the processes of nuclear fusion are exhausted. Just what they settle into depends on their mass.

A dying star, on this picture, consists of a dense ball of matter, chiefly in the form of nuclei of iron swimming in a sea of free electrons. Of course, there will still be a thin atmosphere of hydrogen and helium, and probably a crust rich in nuclei such as those of carbon, but we can ignore those. The most important particles, as far as the first stage of stardeath are concerned, turn out to be the electrons.

Electrons have a very important property, which they share with other entities that we are used to thinking of as particles, such as protons and neutrons. This family of particles is known as fermions, after the Italian physicist Enrico Fermi, and no two fermions can ever share exactly the same quantum 'state'. This is the reason why the electrons in an atom, for example, are spaced out around the nucleus. When physicists first discovered that electrons in atoms formed a cloud around the nucleus, they were greatly puzzled, because the electrons have negative charge and the nucleus positive charge, and opposite charges, of course, attract. Why didn't the electrons all fall into the nucleus? In essence, the answer is that if they did so they would all be in the same state, in the same energy level. In order to keep its own uniqueness, each electron associated with an atomic nucleus has to find its own place, somewhere near the nucleus, jostling with the other electrons that belong to that atom in a cloud surrounding the nucleus.

There are 'particles' which are not fermions, and which are called bosons, after another pioneering physicist, the Indian Satyendra Bose.

These are the particles that we often think of as waves, or radiation. Photons, for example, are bosons, and they are quite happy to crowd into the same state as one another. Indeed, this is the basic principle of the laser – the powerful beam of light which a laser emits is made up of countless numbers of photons, all in exactly the same state, all marching in step with one another. But that has nothing to do with what holds a dead star up against the pull of gravity.

As a dead star cools and shrinks, there comes a time when all of the electrons are packed so closely together that they occupy all of the possible states that exist for them inside the star. They can be squeezed together no more, since that would involve forcing several electrons into the same state, and they resist the inward tug of gravity with an outward pressure, called electron degeneracy pressure. Provided the mass of the stellar remnant is less than about 1.4 times the mass of our Sun, this is the end of the story. Gravitational force balances the electron degeneracy pressure when a star like the Sun has shrunk to about the size of the Earth, and it is called a white dwarf. Astronomers know many white dwarfs, and, indeed, such stars were discovered before the theory I have outlined here was put together to explain their origins. As such a star cools, it stays much the same size, but gets dimmer and dimmer, fading through the brown dwarf stage to become, ultimately, a black dwarf – a ball of iron, surrounded by a shell of carbon, and perhaps even a trace of ice from the remnants of oxygen and hydrogen in its old atmosphere. The fate of our Sun is to become a sooty ball of ice-streaked rust.

But what if the mass of the star is more than 1.4 times the mass of our Sun, even after it has blown away its outer layers in a last blaze of glory? Although the electrons still cannot occupy the same state as each other, under such intense gravitational pressure they find another escape route, as, in effect, electrons are squeezed into protons to form neutrons, releasing neutrinos in the process. It might seem that they would be happy to do so, since positive and negative charges are attracted to one another, but in fact there are other forces at work in atomic nuclei (and under comparable conditions), which keep protons and electrons apart unless the squeeze is overwhelming. I won't go into the details here; what matters is that if a dead star has got more than 1.4 times the solar mass of material doing the gravitational squeezing, all the electrons and protons it contains are converted into neutrons, and it collapses a stage further, to become a ball of neutrons – in effect, a single 'atomic' nucleus – about 10 kilometres in radius. It has become a neutron star. At this point, it is held up by the degeneracy pressure of the neutrons, exactly similar to the electron degeneracy pressure which holds a white dwarf up against the pull of gravity. But this pressure can only resist gravity if the mass of the stellar remnant is less than about 3 times the mass of our Sun.

Some physicists have suggested that there may be a stage beyond the neutron star stage, in which the neutrons are reduced to their ultimate components, quarks, and the star is made of quark 'soup'. But this has very little effect on the calculations, since quark soup is just about the same density as neutron matter, anyway. With no nuclear fuel to burn, any dead star with a mass of more than 3 times the mass of our Sun cannot hold out against gravity at all, but must collapse into the ultimate sink, a black hole, while the matter it once contained is literally crushed out of existence. Neutron stars and black holes may very well be made in the explosions that produce supernovae; this would also be a way to make neutron stars with less than 1.4 solar masses, by squeezing their material together at the hearts of exploding stars. Black holes will also be formed if another compact object – a neutron star, perhaps – sweeps up enough matter from its surroundings to exceed the 3 solar mass limit. Whatever their origin, there are plenty of these compact objects about. Hundreds of neutron stars have been detected, in the form of pulsars, and several likely black holes are known. Every star *must* end up as a white dwarf, a neutron star, or a black hole. But what happens then?

The fate of matter

In round terms, the age of the Universe today (the time that has elapsed since the Big Bang) is about 10^{10} years. Cosmologists can calculate how matter will evolve as time passes in model universes with different densities. The best bet, as we have seen, is that our Universe sits very close to the dividing line between being open and closed, with omega very close to 1. In constructing such cosmological

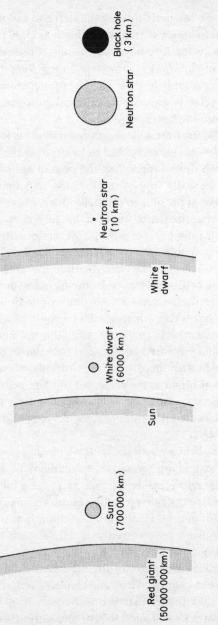

Figure 11.1 Sizes of Stars

The range of possible stellar sizes, shown to scale. Planet Earth is roughly the same size as a white dwarf; a neutron star is the size of a mountain.

models you can make the cycle time of the universe as long as you like, by making omega bigger than 1 but setting it as close to 1 as you like, without ever allowing it to be exactly 1 – or, of course, you can make the lifetime of the model universe *infinitely* long, by setting omega just below (or a lot below!) 1. The fate of matter depends on how long the Universe is around for different processes to work themselves out before the big crunch.

Just about the smallest time scale worth considering is for a closed universe which begins to contract about 10^{11} years after the Big Bang. That is, in a time ten times longer than the present age of our own Universe. Nobody seriously suggests that we live in a universe that small, but it is just about possible to reconcile observations of the real Universe with the requirements imposed by such a model. In a universe this small, the same sort of processes of star formation, planet formation and – presumably – the emergence of life will still be going on in galaxies like our own even after the time the universe turns around and begins to collapse. The first landmark event after the turn around will be when the universe has shrunk to one-hundredth of the size of our Universe today, when galaxies begin to merge into one another. Life as we know it might be around even under those conditions, and would measure a background radiation of only about 100 K. At a size one-thousandth of that of the Universe today, however, the sky has become as bright as the surface of our Sun because of the blueshift effect, piling up radiation from earlier epochs; the 'background' radiation has a temperature of 1,000 degrees, and life as we know it is impossible.

At one-millionth of the present size of the Universe, stars explode as the background temperature reaches several million degrees, comparable to the temperature inside the Sun today; at a billionth the present size, nuclei are broken up into protons and neutrons at a temperature of a billion degrees; and at one-trillionth of the present size protons and neutrons are smashed apart into a quark soup, with a temperature of about a trillion (10^{12}) degrees.

It is certainly a dramatic image of the big crunch,[2] but one that should be treated with caution. First, our Universe contains ten (or more) times more dark matter than it does baryons, and we cannot be sure how these particles will affect the baryons during the collapse.

Secondly, there is no evidence that our Universe, even if closed, is small enough for this scenario to play itself out. Indeed, all the evidence is that our Universe is either just open or (the interpretation I find most persuasive) only just closed. In either case, there will be ample time available for very long-term, quantum effects to come into play. And that gives matter as we know it a very different ending.

If the expansion of the Universe continues for long enough, star formation will cease as all but a trace of the available hydrogen and helium is used up. From studies of the distribution of old and young stars in our Galaxy, and calculations of the rate at which star-making material is being used up, astrophysicists estimate that this will happen in about a trillion (10^{12}) years from now. Galaxies will become redder, as their stars age and cool, and then, eventually, fade away as all the stars they contain become white (ultimately black) dwarfs, neutron stars or black holes. Over very long time scales, galaxies will shrink. This is partly because they lose energy through gravitational radiation, and partly a result of inevitable encounters between stars in which one star gains energy and is ejected from the galaxy while the other loses energy and falls towards the centre of the galaxy. In a similar fashion, clusters of galaxies will shrink in upon themselves, and eventually both individual galaxies and clusters will fall into huge black holes of their own creation.

It is difficult to come to grips with what 'eventually' means. The numbers cosmologists play with look simple enough – 10^{15} years, 10^{20} years, and so on. But remember that each additional unit on the 'power of ten' means *multiplying* tenfold. The age of the Universe is about 10^{10} years; 10^{11} years is *ten times longer* than all the time that has elapsed since the Big Bang. Similarly, a trillion years (10^{12}) seems mind-bogglingly long to us, but 10^{15} years is a *thousand* times longer still; and 10^{20} years isn't twice as long as the age of the Universe, but ten billion times as long! Even this, however, is a mere eyeblink compared with the time scales required for the ultimate decay of matter.

According to some of the most favoured present-day theories of particle physics (which are favoured because they have been proved right so many times already), protons themselves should be unstable, and must decay, each one turning into a positron, a shower of neutrinos and gamma rays (neutrons inside a white dwarf or neutron star do much the same, but produce an electron as well as a positron, keeping the balance of electric charge). The same rules which allow baryons to be produced in the Big Bang suggest that they must ultimately leave the cosmic stage, with the aid of those tiny deviations from perfection in the C, P and T rules. But the time scale required for this process is very long indeed. In a lump of matter (any lump of baryons), half of the protons will decay in rather more than 10^{31} years. For everyday purposes, the proton is very stable – which is just as well, or we wouldn't be here. But for a slowly cooling white dwarf, proton decay becomes important. Without proton decay, a white dwarf will radiate all its heat away and become a black dwarf at the same temperature as the background radiation in about 10^{20} years. Protons decaying inside it, however, can provide enough energy to keep it as warm as 5 K (a mere $-268°C$) until 10^{31} years have passed. That doesn't sound too hot, but at that time the temperature of the cosmic background radiation will be only 10^{-13} K, which puts it in a more impressive perspective. Neutron stars, being more compact, are kept hotter, perhaps as warm as 100 K, for the same time. By then, half the baryons have been used up.[3] But by 10^{32} years, virtually *all* the baryons have gone, because of course, 10^{32} is *ten times* more than 10^{31}.

By this time, all objects made of baryons have lost almost all their mass. Black dwarf stars have shrunk to the mass of the Earth, and a planet like Earth will have shrunk to the size of an asteroid. By the time 10^{33} years have passed, essentially all the baryons in the universe will have gone, converted into energy, neutrinos, electrons and positrons. Today, every proton in the Universe is balanced by an electron, so that there is no overall electric charge. After this enormous time has passed, and baryons have decayed, the remaining form of matter will be equal numbers of electrons and positrons, scattered across the universe. When lumps of matter, like stars, decay in this way, the positrons and electrons will quickly meet one another and annihilate, releasing more energy in the form of gamma rays. But perhaps as much as 1 per cent of all the original baryons in the universe will still be in the form of hydrogen gas after star formation has stopped, and when the nuclei of these isolated atoms of hydrogen decay the resulting positrons can pair up with the electrons of the original atoms, orbiting one another at a safe distance in a kind of pseudo-atom called positronium.

After 10^{34} years, the universe will contain nothing but radiation, black holes and positronium. Even a black hole, however, is not forever. Stephen Hawking has shown that quantum effects will slowly convert black holes themselves into particles and radiation as they 'evaporate'. A black hole with the mass of a galaxy will evaporate in 10^{99} years, and even a hole containing the mass of a supercluster of galaxies – the biggest likely to form – will be gone in 10^{117} years. The ultimate products of this evaporation will be more electrons and positrons, more neutrinos, and more gamma ray photons. So, after 10^{118} years, if the universe lasts that long, we come to the ultimate fate of matter – to be converted into positronium, neutrinos and photons. And if the GUTs are wrong and protons do not decay as expected, this only shifts the time scale a little matter of four powers of ten, since even a proton will evaporate through the Hawking process after 10^{122} years!

In a closed, but long-lived, universe, recollapse eventually occurs and the plummet towards the omega point still happens, but there are no stars and galaxies to be disrupted in the process. Just a broth of electrons, positrons, neutrinos and photons being squeezed into the ultimate singularity. Somewhere in between the two extremes, medium-sized closed universes will recollapse while they still contain a mixture of dead stars in one of their three guises and some gas and dust that has not been processed into stars.

It is pure guesswork how the dark matter will affect these calculations, since we do not know what the dark matter is. Even so, it seems that those who like to speculate about these things have a choice of futures to 'believe in', and can pick the one that they find most comforting. But this is still not quite the end of the story, since there is another sense in which we seem to have a choice of universes, an interpretation of cosmology which brings us, like a closed universe, round full circle and back to the more philosophical considerations with which this book began.

The moment of creation

It is quite astonishing how old ideas keep turning up in new guises as we probe back to the moment of creation. The Steady State model in its simplest modern form, originally put forward in 1948, is long since discredited, because we can see that the Universe is evolving, changing as it expands away from a superdense state. But in the 1960s, fighting a rearguard action against the advance of Big Bang cosmology, Fred Hoyle and his Indian colleague Jayant Narlikar developed a variation on the theme. This involved an eternal Steady State universe in a state of high temperature and density, within which bubbles might sometimes be inflated to produce regions of expanding, evolving spacetime just like the Universe we see about us; a curious preview of the kind of inflationary model now so fashionable.

In order to provide a driving force behind this inflation, Hoyle and Narlikar had to invent a new kind of field, which they called the C field, because they designed it to be involved with the creation of matter. So they had a model of the Universe, more than twenty years ago, in which a very dense state was driven into a burst of rapid expansion by the energy of a field which also made particles. Conceptually, there is very little difference here from the now fashionable idea of the inflation of the Universe out of a very dense state by the action of a field or fields which dumps its energy in the form of particles. In 1984, Hoyle and Narlikar each separately published papers drawing attention to the similarities between their old C-field model and inflation. The point they hoped (in vain, so far) other cosmologists might pick up from this idea is that the inflationary process, like the C field, can operate quite happily against a steady state background. Physicists have not, to be frank, fallen upon this revival of the C-field idea with cries of delight. *Any* variation on the Steady State idea is unfashionable. But credit where credit is due – the mathematics of the C field *is* reminiscent of inflation (or, as Hoyle and Narlikar would say, inflation is reminiscent of C-field theory), and it *is* important to keep in mind the possibility that there may have been no actual singularity at what we call the moment of creation, no state of literally infinite density and temperature, but that there may have been instead a transition of a local region of spacetime, which we call the Universe, from one state to another. More of this shortly.

But one of the most dramatic implications of the idea of inflation in its 1985 form is that the whole Universe may have appeared out of literally nothing at all, created as a quantum fluctuation in the same way that quantum uncertainty allows a virtual pair of particles to appear and to exist for a short time before annihilating.

The idea surfaced in *Nature* in December 1973, in the form of a scientific paper from Edward Tryon of Hunter College, City University of New York. Tryon proposed[4] what he called 'the simplest and most appealing' Big Bang model imaginable, that 'our Universe is a fluctuation of the vacuum'. The jumping-off point for his introduction of this model into cosmological debate was a calculation which showed that any closed Universe must have zero net energy.

Crudely speaking, we can understand this, as I have already hinted, in terms of the negative gravitational energy that the Universe possesses, which is so large (in a negative sense) that it cancels out all of the mass energy, mc^2, of the matter in the Universe. This is a consequence of a feature of gravity so strange that physicists scarcely seem to acknowledge it in public. If we try to describe the gravitational energy of a collection of matter in terms of the equations that describe gravity, we find that the only meaningful state which corresponds to a zero of energy from which we can measure is with all the matter dispersed to infinity. It doesn't matter if we think in terms of atoms, or planets, or stars, or galaxies as the building blocks of the material object under consideration. The zero of energy is when the building blocks are as far apart as it is possible to conceive. Now comes the strange thing. As the matter falls together under the influence of gravity, it gives up energy, so the gravitational potential energy gets less, and since it was zero to start with that means it becomes negative. So any object in the real Universe, like a planet, which is *not* spread out to infinity must have a negative energy to start with, and if it shrinks it releases energy and its own gravitational potential energy becomes more negative. Not something for nothing, but something for *less* than nothing. Using Newton's theory of gravity, there is no limit to how negative the potential energy of the planet can get. Every time it shrinks, energy is released and the gravitational energy becomes more negative.

In mathematical terminology, there is no 'lower bound' to the energy state, in Newtonian physics, and that is one reason why we have to set zero as the energy of the dispersed matter. General Relativity does set a bound to the amount of negative gravitational energy associated with a body with mass m. If all of the mass m could be concentrated at a point in space, a singularity, then the negative gravitational potential energy associated with it would be $-mc^2$ – equal and opposite to its Einsteinian mass-energy. But this is only a crude representation of a more subtle and sophisticated mathematical argument, which I can't go into here, which *proves* that a closed Universe has zero energy overall.[5] It is Guth's 'free lunch' taken to its logical extreme; if the Universe contains zero energy, no wonder it is free. Not something for nothing, after all, but *nothing* for nothing. Tryon pointed out that the uncertainty relation $\Delta E \Delta t = \hbar$, allows anything with zero energy to exist for as long as you like, because if the energy is zero then the uncertainty in the energy ΔE, is also zero. There would be no problem about 'borrowing' energy from the vacuum to create the Universe, because you don't need any overall energy in the first place, and you don't have to hurry to pay it back, because there is nothing missing from the balance account!

This naïvely simplistic interpretation of the uncertainty rules made no great splash in the 1970s. It was just one of those passing comments that physicists often toss around during coffee time, and it clearly did not provide a precise description of our Universe. Taking the analogy with the creation of pairs of virtual particles literally, it would require, for example, that our Universe contained precisely equal amounts of matter and antimatter, which doesn't seem to be the case. And the whole basis of the argument was that the Universe is closed, whereas in the mid-1970s the well-established consensus among cosmologists was that the Universe was open. Finally, if a quantum fluctuation containing all of the mass of our Universe *were* created in a superdense state, why on Earth didn't it promptly collapse into a singularity under the influence of its own self-gravity?

The difficulties looked insurmountable. But the advent of inflation changed all that. Inflation requires that there must be enough dark matter to make the Universe so nearly flat that we cannot distinguish its curvature, and inflation doesn't 'care' whether the Universe sits just on the closed side of flatness or just on the open side of the dividing line. And the GUT description of X-boson decay tells us that an energetic original universe created with equal numbers of X and anti-X will still evolve into a Universe with a slight residue of matter in it, in the fullness of time (the 'fullness of time' being about 10^{-35} second). Hardly surprisingly, Ed Tryon himself revived his idea in the context of inflation in the 1980s, and in 1982 it was also taken up by Alexander Vilenkin, of Tufts University.[6] Vilenkin, indeed, takes things a step further than Tryon did in 1973. Tryon talked about a 'vacuum fluctuation', implying that some form of spacetime metric existed before the Universe came into being; but Vilenkin is trying to develop a model in which space, time and matter are all created out of literally nothing at all, as a quantum fluctuation of nothing. 'The concept of the universe being created from nothing is a crazy one,' Vilenkin says in one of his papers (*Physics Letters*, Volume 117B, page 26, 4 November 1982); but he goes on to show how it is mathematically equivalent to the creation of an electron-positron pair that then annihilate each other, and which is in turn equivalent to one electron being created out of nothing, travelling forward in time for a while, then turning round and travelling backwards in time to meet up with its own creation. In this and other papers, Vilenkin puts a lot of respectable mathematical icing on to the basic cake baked up by Tryon, in more speculative form, in the early 1970s.

Hardly surprisingly, this revival of his speculation has meant a great deal to Tryon, especially since he has trodden a lone path in all his cosmological work. Born in 1940, in Terre Haute, Indiana, he obtained his Bachelor's degree in physics from Cornell in 1962, then moved to the Berkeley campus of the University of California, where he was captivated by Steven Weinberg's courses in quantum field theory and General Relativity, and was lucky (and talented) enough to become one of Weinberg's PhD students. After finishing his thesis, he then moved to Columbia University, carrying out calculations of the scattering amplitudes for collisions between pairs of pions, and earning the nickname 'Pion Tryon' in the process. But he has never collaborated with anybody, and, possibly uniquely among his generation of particle theorists, is the sole author of all of his scientific publications. An interest in cosmology that went back to childhood had already led Tryon to speculate about the possibility of a closed Universe with zero net energy, but he recalls with amusement an incident at the end of the 1960s which, at the time, caused him acute embarrassment.

Cosmologist Dennis Sciama was visiting Columbia from Britain, and gave a seminar on the latest theories of the Universe. At a point in the presentation where Sciama paused for a moment, Tryon blurted out, to his own surprise as much as everyone else's, 'maybe the Universe is a vacuum fluctuation!' The laughter that followed caused the junior researcher, who specialized in particle physics, not cosmology, to blot the incident from his mind. It was only after the *Nature* paper appeared that he was reminded of the occasion by a colleague, and realized that his subconscious had been prodding him towards completing that piece of work, even though the memory had been kept from his conscious mind for three years.

Tryon moved to Hunter College in 1971, and it was there that the image of the Universe as a quantum fluctuation appeared to him in mid-1972, in a flash, fully worked out. The subconscious had been absorbing all his reading on cosmology and working out the answer to that embarrassing laughter, only releasing it into his conscious mind when it was complete. And when it was published in *Nature*, the article drew some 150 requests for copies, even though the idea then fell into limbo until the concept of inflation made it timely. How did it feel to have an old idea, once literally laughed to scorn, made fashionable and referred to as a major conceptual advance? 'All good scientists are dreamers,' said Tryon when I asked him about this. 'They dream of discovering some unknown phenomenon of major importance ... it would be difficult to exaggerate the satisfaction entailed.'

Of course, the idea is still highly speculative, but it is now much more attractive. And its best feature is that there is no longer any need to create all of the matter in the present-day Universe at the moment of creation itself. All you need now is to create a region of closed spacetime and energy, a self-contained microcosm much smaller than a proton, which has only a modest temperature and a slight tendency to expand. Without inflation, such a microcosm would soon collapse. But with inflation, says Tryon, there could have been a 'cold big whoosh' (inflation) which blew the tiny speck of spacetime up to an enormous size and ended in a burst of creation as the quantum field energy was dumped into pairs of Xs and other particles and created the hot Big Bang itself at $t = 10^{-35}$ second. Ever since, gravity has been at work to slow down the expansion, and eventually it will be first halted and then reversed. In the far distant future, the Universe will collapse back into a tiny singularity. Spacetime itself, and everything it contains, will disappear into a single point and vanish. There will be nothing to show that our Universe ever existed. The proposal may remain speculative, but at least it now fits rather neatly into the overall framework of inflationary cosmology. If there really was a moment of creation, then the concept of the quantum uncertainty, one of the strangest and most fundamental features of quantum physics, seems to provide the best hope of explaining how the Universe came into being. In that case, there may indeed have been a moment of creation, marking the boundary of the Universe at the beginning of time. But there is an alternative view, one that is, if anything, even more deeply rooted in the basics of quantum physics. Stephen Hawking, of the University of Cambridge, has developed an approach based on the concept of defining a quantum mechanical wave function that describes the entire Universe, and dealing with this, as one could any other wave function in quantum physics, in terms of path integrals. And he says that there may be no boundary to the Universe, even at the moment of creation.

A seeker of singularities

Stephen Hawking, now widely known as the author of a best-selling book about the Universe, is one of those rare scientists whose work captures the popular imagination and is often the subject of reports in newspapers and magazines. This is partly because he works on topics that seem to strike a chord with all of us – black holes, singularities in spacetime, and the mystery of the origin of the Universe. But it is also because he suffers from a severely disabling disease, called motor neurone disease. This is the disease which killed the actor David Niven; it attacks the nervous and muscular functions of the body, making it impossible for a severely afflicted sufferer – which Hawking now is – to walk, and extraordinarily difficult for him to talk. The image of a crippled genius struggling against overwhelming odds to achieve an understanding of the Universe far beyond most of us, and then struggling to communicate those ideas to his colleagues in spite of his handicaps, is one that obviously makes for 'good copy', in newspaper parlance. But sometimes too much is made of this aspect of Hawking's life. Certainly his body is crippled. But his mind and intellect are totally unaffected by the disease, and his scientific achievements are of the first rank in themselves. It isn't that it is surprising that someone with his physical problems can make such progress in understanding the Universe; what is surprising is that any human being can achieve such understanding. And Hawking himself has commented on his great good fortune, as he sees it, of having a career which depends solely upon the ability of his brain to think, and in which his physical handicaps, severe though they are, rank as only minor distractions in his own eyes. Of course, these problems are of much greater significance to his life outside his work – but that is about as relevant to his scientific discoveries as the fact that Einstein played the violin is to the story of the search for the Big Bang.

Hawking was born on 8 January 1942 – precisely, as he delights in telling people, three hundred years to the day after Galileo died. His father worked on research into tropical diseases at the National Institute of Medical Research, and encouraged Stephen to follow an academic path aimed at entrance to Oxford University. The encouragement did not, however, extend to Hawking's decision to study mathematics – his father tried to talk him out of this, arguing that there were no jobs for mathematicians. Even so, Stephen entered Oxford University in 1959 to study mathematics and physics. His contemporaries and tutors recall today that he was a remarkable student with a mind unlike that of anyone else. He passed examinations with almost contemptuous ease, obtained a First Class degree, and moved to Cambridge to begin research on cosmology.

At this time, in the early 1960s, Hawking began his involvement with singularities, an involvement that lies at the heart of all his major contributions to science, and which is the key to understanding the moment of creation itself. He was, and is, fascinated by the idea of a mathematical singularity, a point where not only matter but space and time as well are either crushed out of existence or, in the case of the Big Bang, created. The standard equations of relativity theory predict the existence of singularities, but in the early 1960s hardly anybody took this prediction seriously. Singularities were assumed to be an indication that the simplest version of Einstein's theory, with a smooth distribution of matter through spacetime, was not a realistic way to describe the confusion of a superdense state, and that a better understanding of the equations would probably show that as a collapsing object approached a singularity, at some stage there would be a 'bounce', making it expand again, or some other effect that halted the collapse short of a point of infinite density. Either that, or Einstein's theory was incomplete, and would break down at very high densities, that is, in very strong gravitational fields. Hawking determined to find out if this were true. But it was to be several years before this determination bore fruit, because it was in his first year of graduate work, 1962, that the first symptoms of his illness appeared and were diagnosed. Given only a few years to live, he became depressed, took to drink, and virtually gave up his work.

But as the months passed it became clear that the progression of the disease had halted and stabilized. Hawking was slightly incapacitated physically, but he wasn't getting any worse. And at the same time he realized first that his intellect had been totally unaffected by the disease, and would be unaffected whatever happened to his body, and also that his work was entirely brain work, which could be carried on regardless of the deterioration of his physical condition. Since then, as far as a casual acquaintance such as myself can tell, Hawking has never looked back, either in his private life or in his work. He married in 1967, has two sons and a daughter, and leads as normal a life as possible. It was also in the late 1960s that he began to achieve recognition for his scientific work.[7]

One of Hawking's major achievements at this time was carried out in collaboration with mathematician Roger Penrose, who was then working in the University of London. Together, they proved that the equations of General Relativity in their classical form (that is, without allowing for quantum effects), absolutely *require* that there was a singularity at the birth of the Universe, a point at which time began. There is no way around the singularity problem within the framework of classical General Relativity. If singularities are to be avoided in the real Universe, the only hope is to improve relativity theory by bringing in the effects of quantum theory and developing a quantum theory of gravity. In the 1970s, Hawking's investigations of the mathematics of black holes led, through the introduction of quantum effects, to the startling conclusion that black holes can 'evaporate' and must eventually explode. This work brought him into the popular limelight, at least in the science magazines. And in 1974, at the very young age of thirty-two, he was elected a Fellow of the Royal Society.

By then he was confined to a wheelchair following a further progression of his illness. He has only very limited control over the muscles of his body, and slumps rather than sits in his wheelchair; he communicates through a computer and voice-synthesizer system mounted on his wheelchair. The honours Hawking has received include the Albert Einstein Award, in 1978,[8] and in 1980 he became Lucasian Professor of Mathematics in the University of Cambridge – a chair occupied previously by Paul Dirac and Isaac Newton, among others. These honours, and the honorary degrees heaped upon Hawking by universities around the world, are the sort of thing usually associated with a scientist who has completed his greatest work and can now settle down to a comfortable position of eminence as an administrator and teacher. Few mathematicians, in particular, achieve much in the way of new work after they reach the age of thirty; new ideas come from young minds that are not hidebound by convention, or so we are told. But Hawking's mind is as sharp as ever, and he has now put forward a model of the Universe which attempts to combine the ideas of General Relativity and quantum physics, and which not only removes the uncomfortable singularity at the moment of creation but which, in principle, explains *everything* in one package. Perhaps his disability is now actually an advantage, as far as this work is concerned. For, unable to become a conventional committee man and figurehead, he continues to work in the only way he can, the way he has worked for twenty years, developing new mathematical descriptions of the world in his head. The astonishing thing about this latest work, which may well be remembered as Hawking's masterpiece, is not that it comes from someone with physical disabilities. The surprise is that it comes from a man who was then in his early forties, a remarkably late age for a mathematician to be achieving a new breakthrough. Even Einstein was still in his thirties when he completed his General Theory, and he never achieved much of any significance in science after that.

The model Hawking proposes, and which is still being developed, has not yet won the kind of acceptance that his earlier work has achieved. But it strikes to the heart of the remaining puzzles about the origin of space, time and matter, and it is quite clearly the most complete, coherent account of the moment of creation that is on offer today.

Quantum realities

General Relativity tells us that there must be a singularity at the beginning of time – the moment of creation. But General Relativity, like all of our theories of physics, breaks down for times earlier than the Planck time, 10^{-43} second. Although a variation on the Steady State idea, with either a smooth meta-universe or some sort of overall chaos, plus inflation, could provide a way to produce a local region of expanding spacetime rather like the one we live in, it would be much more satisfying if we could develop a mathematical model, a set of equations, to describe our Universe in a self-contained way – especially if that model could avoid the embarrassment of a singularity at $t = 0$. This is the basis of Hawking's approach to the puzzle of our origins, and of his attempts to combine General Relativity and quantum theory, at least partially, in a good working model of the Universe.[9]

What version of quantum physics, however, is appropriate when we are describing the whole Universe? Remember that quantum theory tells us nothing about how a particle, or a system, gets from state A to state B. The conventional interpretation of quantum physics is the Copenhagen Interpretation. One simple example will help to highlight its oddity.

Every electron carries a property which physicists choose to call spin. You can think of it, if you like, as an arrow, or a spear, which is carried by an electron and can only point in one of two directions, 'up' or 'down'. Really, we shouldn't try to think of these quantum properties in everyday terms at all, but it is the only way to get any kind of a picture of what is going on. Whatever the image you carry in your head, though, what matters is that an electron with spin up is in a different state from an electron with spin down. This is why two electrons can both crowd into the lowest energy level available in the helium atom, for example, without being in the same state. Because one has spin up and one spin down they can share the energy level, in a sense both at the same distance from the atom's nucleus. If they both had the same spin, this would be excluded, because they are fermions. In a more complex atom, the next electron has to go into a higher energy level, further out from the nucleus, because whichever possible spin it has it is excluded from the lowest level by the presence of another electron with the same spin – but that requirement, with all that it implies for chemistry, is not the story I want to go into here. Instead, imagine a single electron, sitting on its own, or travelling through space. What spin does it have?

Now, we can do experiments to measure the spin of the electron, and when we do we will always find that it either has spin up or it has spin down. But the quantum theory tells us that when the electron is left to its own devices, it *neither* has spin up *nor* spin down, but exists instead in some mixture of both possibilities, called a superposition of states. The 'reality' of an electron which has 'collapsed' into a single, definite spin state exists only when it is being measured, or when the electron is interacting with another particle. Once the measurement (or interaction) has ceased, the unique spin state dissolves away once more into a superposition of states. At this level, things only have a unique, 'real' existence when they are being looked at, or prodded in some way. And this bizarre behaviour holds true for *all* quantum properties, not just spin; it is an *essential* feature of quantum physics. Without it, we would not be able to explain how lasers work, or why DNA forms a stable double helix, or how semiconductor chips do their tricks inside computers, to name but a few. As for what this strange behaviour implies about the nature of reality, physicists and philosophers have been debating the issue, on and off, for half a century. Now, the cosmologists have got in on the act.

There are two principal ideas about what the collapse of an electron (or anything else) from a superposition of states really means. The conventional wisdom is called the Copenhagen Interpretation because a lot of pioneering work on quantum theory was done at the Institute founded by Niels Bohr in Copenhagen. In fact, a key ingredient of this interpretation actually came from Germany, from the work of Max Born. This is the idea that the behaviour of things at the quantum level is ruled by chance – not in the whimsical, fluky sense that we sometimes imagine our lives to be ruled by chance, but in the sense that the behaviour of electrons and the like follows the strict statistical rules of probability – the rules a casino takes good care to apply to make sure it keeps an edge. In terms of our isolated electron, this means that when its spin state is measured there will be a precise 50:50 chance of finding it with spin up, and the same chance of finding it with spin down. Measure a million electrons, or the same electron a million times, and half a million times you will get the 'answer' spin up, half a million times you will be told spin down. But you can never predict in advance what the outcome of any one of the individual measurements will be, only the relative probabilities of all the possible different outcomes. The same thing happens when you toss a coin. It has a 50:50 chance of coming up heads (assuming the coin is properly balanced), and each time you toss the chance is the same, even if you have just had a string of heads or tails. For an individual electron, even if you measure its spin and get the answer 'up', there is only a 50:50 chance that the next measurement will give the same answer. It is only because real-life experiments involve huge numbers of quantum objects, all following the rules of probability, that their *overall* behaviour can be predicted, for practical purposes, using statistics. Half the electrons have spin up, half spin down, and a TV tube, for example, doesn't care which half are oriented which way. The quantum rules of probability work rather like the way in which life insurance companies make their money – they cannot tell in advance *which* of their individual clients will die in any particular way, but they do know from their actuarial (statistical) tables, *how many* will die, and budget accordingly.

The choice of spin states for a single electron is a very simple example, and real quantum systems are much more complicated superpositions of states, governed by suitably more tricky rules of probability. The probabilities themselves may be changed by the measurement, or interaction. For example, one of the other strange features of quantum physics is that an object like an electron is not confined to a definite location. It has a certain probability, which can be calculated, of turning up anywhere at all (this is related to the wave-like aspect of the electron's dual nature). The probability is very large that the electron will be found somewhere near where you last saw it, but there is a real, if tiny, probability that it will turn up somewhere else entirely. When you actually do measure the position of an electron, these probabilities all collapse into the certainty that it is where you saw it. Once you stop looking, however, it once again has an opportunity to be somewhere else. Looking at the electron changes its probabilities, and alters the superposition of states the electron entity is in.

It sounds crazy. But the strangest thing about the Copenhagen Interpretation is that it works perfectly as a tool to describe what will happen as a result of our interference with the quantum world. It is the tool used by people working at the practical level, and it has proved its practical value repeatedly. The equations work. But nobody has the faintest idea what the interpretation really means, in terms of what electrons and things 'do' when nobody is looking at them. This has provided scope for an alternative interpretation to be suggested, one which gives exactly the same 'answers' as the Copenhagen Interpretation in all practical applications, but which has a different philosophical basis. It stems from the work of the American Hugh Everett, in the 1950s; for reasons which will soon be apparent, this is called the 'Many Worlds Interpretation'.

Quantum cosmology

The Many Worlds version of quantum theory says that when you measure the spin of an electron it does *not* collapse into one spin state (chosen by the laws of probability) while you are looking at it, and then revert to a superposition of states. Instead, according to this interpretation, the world splits into two separate realities, in one of which the electron has spin up, and in the other of which the electron has spin down. The two worlds then go their separate ways, nevermore to interact with one another. Physicists and mathematicians are still arguing about exactly what this means – especially when scaled up to cover systems with more complex superpositions of states than an electron with only a choice of two spins. The fundamental thing to grasp, though, is that calculations carried through on the basis of the Many Worlds Interpretation give exactly the same answers as the Copenhagen Interpretation, always, when applied to practical problems. So it is just as good a tool (no better, no worse) for designing computers, or lasers, or in calculating the chemistry of complex molecules. The implications though, are more than interesting.

One idea is that the whole universe splits into two or more replicas of itself every time any quantum system is forced to choose between possible states. This has led philosophers to fling up their hands in horror, unwilling to accept the implication that my measurement of the spin of an electron, here in a lab on Earth, can affect galaxies and quasars millions of light years away, instantaneously, as the entire universe splits in two. Or, indeed, that every time a quantum 'choice' is made in some distant quasar, you, and I, and everyone here on Earth is duplicated into a myriad copies. (Science fiction writers, of course, love the idea!) But there is a way to appease the philosophers. Researchers such as Frank Tipler, of Tulane University, prefer to associate the 'splitting' with the quantum system involved in the interaction, or the experimental apparatus involved in making the quantum measurement. 'There is only one Universe,' says Tipler, 'but small parts of it – measuring apparata – split into several pieces.'[10]

The implications of applying Many Worlds theory to cosmology are so startling that it may yet supplant the Copenhagen Interpretation, just as, in the seventeenth century, the Copernican interpretation of planetary motions in terms of the Earth going round the Sun supplanted the Earth-centred ideas of Ptolemy. It also rounds out the theme of this book beautifully. The idea is to take the equations describing the growth of the Universe from the Big Bang and to treat them in quantum terms using the Many Worlds Interpretation. On this picture, the Universe is split into many different branches by quantum processes in the beginning, at the moment of creation when the 'size' of the Universe, as far as that has any meaning, is the size of a quantum fluctuation. These branches will have different properties from one another, but will be part of a single family, governed by a single set of rules. From the point of view of the present book, the most important feature is that there will be branches which have all permissible values of omega. That might seem like a disadvantage of the interpretation, since it seems to beg the question of why omega is so close to 1 in our Universe. But it turns out that this special value of omega is a *requirement* of Many Worlds cosmology.

Hawking's universe

We can imagine the Universe as being described by quantum mechanical wave functions, of course, even if we can never hope to write down the equations that would describe the 'wave function' of the entire Universe. But since, by definition, the Universe includes everything of which we can have knowledge, including ourselves, there is nobody 'outside' the Universe to observe it and thereby to cause it to collapse into one possible quantum state. The correct way to calculate the probabilities that describe the behaviour of the Universe, says Hawking, is to use the Many Worlds Interpretation, in which the effects of all of the possible wave functions for the system can, in principle, be calculated and added together, to produce an overall mathematical description of the system and how it gets from state A to state B.

In most cases, these two approaches give the same answers to the problems of quantum physics, and few physicists bother with the Many Worlds Interpretation because they have grown up with the Copenhagen Interpretation and, having got used to one idea that conflicts with common sense, they find it hard to adjust to a second non-commonsensical idea. But as Hawking has pointed out on several occasions, a combination of the Many Worlds Interpretation and what is called a 'sum over histories' is the *only* way to approach a quantum description of the Universe. The 'sum over histories' is literally that, an *adding together* of all of those possible ways in which the Universe could evolve. Of course, we cannot even calculate one 'history' of the Universe in detail. But we can choose a set of starting conditions for the Universe – boundary conditions – and we (or rather Hawking and his colleagues) can calculate the evolution of a simple version of the Universe, that contains just a couple of quantum fields (one representing gravity, one representing matter). The hope is that this simple model, which Hawking calls minisuperspace, will bear enough resemblance to the real Universe for him to deduce the broad features of the evolution of the Universe. And that hope seems to be fulfilled.

Hawking chooses as his boundary condition for the Universe a possibility which only arises when quantum physics is combined with General Relativity. General Relativity says that there must be a singularity at $t = 0$. Quantum mechanics offers the prospect of removing the singularity at $t = 0$ from cosmology. In physical terms, we can think of the origin of time as being smeared out by quantum uncertainty, over a time of 10^{-43} second, so that there is no unique moment of creation. In terms of a physical model, described by a proper mathematical combination of GR and quantum physics, this makes it possible to describe the *four* dimensions of spacetime as a closed surface like the surface of a sphere, or the surface of the Earth.

Previously, we met the idea of the *three* dimensions of space forming such a closed surface, which expands as time passes. But now we have to think of spacetime, not just space, in this way. Extending the analogy, space would be represented not by a surface but by a line, which we can choose to be a line of latitude circling around the spherical surface that represents the fabric of spacetime. Time could then be represented by the 'distance' from the pole along a line of longitude, and if we start out from, say, the North Pole and move towards the equator the great circles that represent space (lines of latitude) get bigger as 'time' passes and we approach the equator. Such a model of the Universe is completely self-contained. There are no edges, and there are no singularities in either space or time. It is the simplest possible geometry that could describe the Universe, and it is a geometry that can only exist because quantum effects change the rules of relativity theory, which on their own insist that there *must* be a singularity at the beginning of time. Hawking stresses that this proposed state of the Universe is just that – a *proposal*. He suggests that the boundary condition of the Universe is that 'it has no boundary' – no edges, no singularities, no beginning or end of either time or space. The astonishing thing is that this simplest of all possible boundary conditions leads him to an entirely plausible description of the Universe, indistinguishable from what we see about us.

Hawking's tests of this model use the sum over histories approach to quantum physics. In principle, the idea is to add up the effects of all the possible histories which satisfy the boundary condition – all the possible universes that are finite in size and have no boundaries. In practice, he has to make many simplifying assumptions, boiling his model universes down to the basic two fields that I mentioned. But when he does this and carries through the path integrals, he finds that most of the histories cancel out. Only a few of the possible histories are reinforced and therefore have a high probability. They form a family of high probability histories, which share several important properties. One is that they expand uniformly in all three space directions; another is that they each expand out to a definite size, then contract back into a state of very high density like the state of our Universe at the Planck time, before expanding once again. And each cycle of expansion and collapse is exactly the same as the one before. The Universe doesn't expand to a bigger or smaller amount in consecutive cycles, but always by exactly the same amount. Even better, the interaction of the two fields in Hawking's models produces an initial phase of very rapid expansion – inflation – before the matter begins to dominate the Universe and causes it to switch over into the kind of sedate expansion we see in the Universe today.

Any one of these allowed universal histories would be a good description of our Universe, as a closed system with no boundaries and no singularities, eternally fated to carry out a cycle of expansion, collapse and expansion. It is easy to see why Hawking is excited by the possibilities thrown up by his model. The Universe, on such a picture, must be just closed, but an era of inflation will have carried it close to flatness, and solved all the usual problems that are solved by inflation. But the model also throws up some strange and wonderful new ideas. There must be other universes in the family of allowed histories, going through their cycles of expansion and collapse in some sense alongside us (next door in superspace). But there is no way we could ever become aware of them, let alone communicate with them. Because of the way quantum physics works, whenever we make measurements or carry out experiments we will get results in line with one quantum state – the wave function that describes 'our' Universe and everything in it, including ourselves. Intelligent beings that occupy a quantum state which corresponds to a second highly probable wave function of the Universe will make their own observations and always get answers appropriate for that wave function. Apart from cancelling out some wave functions and reinforcing others, the quantum states do not interact – there is no interference, and the results of experiments are always in line with one or the other 'classical' solution to the equations.

Look again at Hawking's model of the four-dimensional Universe as a smooth sphere. The rings of constant latitude that expand outwards from the North Pole represent the expanding Universe of space, and the North Pole itself represents the Big Bang – the moment of creation. But there is no singularity at the pole. It is just a place where we measure time from. In the same way, the fact that the real North Pole of the Earth is a place we can measure latitude from (in fact, we define the *equator* as latitude 0°, but we could just as easily measure from the pole) doesn't mean that there is a singularity at the pole. There is nothing further north than the pole, but that doesn't mean that space has an edge there. And there is nothing earlier than $t = 0$, but that doesn't mean that time began then. The moment of creation, $t = 0$, is now just a convenient label against which to measure time.

Hawking first presented these ideas at a conference on cosmology that was held in the Vatican in 1981. The physicists and mathematicians who attended that conference were granted an audience by the Pope, who told them that it was quite in order for them to study the evolution of the Universe after the moment of creation, but that the puzzle of the beginning of time itself was a matter for religion, not science, and represented the work of God.[11] Perhaps the Pope's advisers had been too tactful to point out to him that Hawking's model of the Universe removed the singularity at the beginning of time, and therefore removed the role assigned to God by the Pope. Or perhaps the full import of what Hawking had told the conference had not sunk in.

Hawking's model removes the embarrassing singularity at the 'beginning' of the Universe, and also the one at the 'end'. It also has important philosophical implications. Even though the details remain to be painted in, and there is still scope for kite-flying speculation, he is telling us that it is possible in principle to develop a mathematical model which describes the Universe completely in terms of the known laws of science alone, without any need to invoke special conditions even at the moment of creation. Quantum physics is the key needed to unlock the last secrets if the Universe and to explain both its beginning and its end.

Our search for the Big Bang, and back before the Big Bang to the moment of creation itself, is over. Hawking's universe holds out the prospect of combining General Relativity and cosmology in one grand theory of creation, and tells us that we already know all of the fundamental laws of physics. There is no need to invoke miracles, or new physics, to explain where the Universe came from. It is now possible to give a good scientific answer to the question 'where do we come from?', without invoking either God or special boundary conditions for the Universe at the moment of creation.

Eternal inflation

But that is still not quite the end of my story. The term 'Universe' is generally accepted to mean everything that we can ever have knowledge of, and Hawking's variation on the quantum theme seems to explain the birth and death of that quite well. But theorists are also able to speculate about the possibility that our bubble of spacetime is embedded in some greater 'meta-universe', which we can never experience or study directly, but which may nevertheless exist, somewhere 'out there' in space and time.

There are several variations on this theme. The one I am going to mention briefly here has been developed by Andrei Linde, and is described in his book *Inflation and Quantum Cosmology*. He points out that an unusual feature of the way the Universe is expanding, with 'empty space' stretching to carry galaxies farther apart, is that beyond a certain distance this expansion is proceeding 'faster than light'. This does not contradict special relativity, because the velocity involved is not that of any signal; it is just the rate at which the general expansion of the Universe separates two distant points. But no signal, of course, can travel faster than light. As a result, everything going on inside a bubble with the appropriate radius cannot influence the bubble next door. Each bubble exists in its own right and expands, not by annexing territory belonging to its neighbours, but by a genuine increase in the available amount of territory. 'Any inflationary domain,' says Linde, 'can be considered as a separate mini-universe, expanding independently of what occurs in the rest of the cosmos.'

But this is still not the end of the story. According to quantum field theory, empty space is not entirely empty. It is filled with quantum fluctuations of all kinds of physical fields. And in the spaces between mini-universes in the expanding meta-universe, the presence of these fields can create new bubbles of inflation. So Hawking's universe may be just one among many. In Linde's words, 'the universe . . . unceasingly reproduces itself and becomes immortal. One mini-universe produces many others, and this process goes on without end, even if some of the mini-universes eventually collapse.' Using the term 'universe' where I would prefer 'meta-universe', he concludes that 'some of its parts appear at different times from singularities, or may die in a singular state. New parts are constantly being created from the spacetime foam . . . but the evolution of the [meta]universe as a whole has no end, and it may have had no beginning.' And that, very definitely, is the end of the story I have to tell about the life and death of the Universe.

Appendix

The End of the Search

For me, the moment when the search for the Big Bang was finally over came in the spring of 1997, when a team I was working with used a disarmingly simple technique, combined with the latest data from the Hubble Space Telescope, to measure the age of the Universe. At a personal level, this resolved a puzzle that had nagged at me for thirty years, since I first encountered it as an MSc student in astronomy at the University of Sussex; at a deeper level, combined with a revision of the estimates of the ages of the oldest stars, based on data obtained by the Hipparcos satellite, our estimate of the age of the Universe made this comfortably older than the oldest stars, dotting a final *i* and crossing a final *t* in the story of the Big Bang.

The puzzle that had worried me since 1967 goes back to Hubble's discovery of the expansion of the Universe at the end of the 1920s, and to the relationship between Hubble's constant (or the Hubble parameter) and both the distance scale and the age of the Universe. Edwin Hubble's early work with Milton Humason suggested that the value of the Hubble parameter was about 500 kilometres per second per Megaparsec. So a galaxy seen to have a redshift of 500 km/sec would be at a distance of 1 Mpc (3.25 million light years), and so on. Now, if you know (or think you know) how far away a galaxy is, and you can measure its angular size on the sky, it is easy to work out how big the galaxy really is – its linear size. The further away a galaxy is, the smaller it looks on the sky, in just the same way that your thumb, at the distance of the length of your outstretched arm, is big enough to cover the Moon, which is really much bigger than your thumb, but is at a distance of just under 400,000 km. Indeed, during a solar eclipse, the Moon itself, which has a diameter of just under 3,500 km, completely obscures the Sun, which has a diameter of 1.4

million km, because the Sun is 150 million km further away from us than the Moon is.

The technique works either way around. If you know how big an object really is, then you can work out its distance from its apparent size (just as surveyors do when they measure the angular size of a standard length rod, using a theodolite, to work out how far away it is). Or, if you know how far away an object really is, you can work out its true linear size from its apparent angular size.

The puzzle is that with a value of the Hubble parameter as high as 500 km/sec/Mpc, the galaxies studied by Hubble and his colleagues would have to be rather close to us. In the same way that the Moon looks the same size in the sky as the Sun, but is really much smaller than the Sun and much closer to us, this would mean that the other galaxies were much smaller than the Milky Way Galaxy in which we live.

It's important to appreciate that this wasn't a ridiculous idea at the beginning of the 1930s. It had been barely ten years since it had been firmly established that many of the 'nebulae' really were galaxies in their own right, beyond the Milky Way. And nobody had more than a vague idea of what our own Galaxy would look like, if we could view it from the outside. It really was a possibility, in 1930, that the Milky Way was the only really large star system in the Universe, and that the other galaxies were much lesser systems. It was only gradually, as astronomers were able to map out the structure of our Milky Way Galaxy (definitively, using radio-astronomy techniques in the 1950s), and discovered that it is a spiral galaxy strikingly similar in overall appearance to the spiral galaxies studied by Hubble and his successors, that the idea that we just happen to live in the biggest galaxy around began to look a little odd. But there was at least one person who saw the puzzle with full clarity at a very early stage – the great astrophysicist Arthur Eddington, whose comments, made in 1933, are strikingly prescient:

According to the present measurements the spiral nebulae, although bearing a general resemblance to our Milky Way system, are distinctly smaller. It has been said that if the spiral nebulae are islands, our own galaxy is a continent ... Frankly, I do not believe it; it would be too much of a coincidence. I think that this relation of the Milky Way to other galaxies is a subject on which more light will be thrown by further observational research, and that ultimately we shall find that there are many galaxies of a size equal to and surpassing our own.[1]

Eddington was using a kind of reasoning now known by a name invented by Alex Vilenkin in 1995 – the Principle of Terrestrial Mediocrity. This says that we live on an ordinary planet, orbiting an ordinary star, in an ordinary galaxy, in an ordinary part of the Universe. In other words, if you were set down at random in any spiral galaxy in the Universe, the view would be very much the same as the view we have from Earth.

The situation in 1933 wasn't quite as clear-cut as Eddington made it sound, because if the Milky Way were by far the biggest galaxy in the Universe, and you were set down at random alongside a star somewhere in the Universe, you would be more likely to be set down in the Milky Way than any other galaxy. The argument from the Principle of Terrestrial Mediocrity really is that either the Milky Way is absolutely dominant, or it is absolutely average.

The other aspect of the puzzle of the large value for the Hubble parameter found by Hubble himself is, of course, that it implies that the Universe is very young. If it has been expanding very rapidly, then it would not have taken very long to reach its present size. But the age implied by a Hubble parameter of 500 km/sec/Mpc is only a couple of billion years, far less than the age of the Earth inferred from geological studies. That is why astronomers were so relieved when Walter Baade's revision of the cosmic distance scale led to a greatly reduced value of the Hubble parameter, implying a correspondingly greater age for the Universe, in the early 1950s.

Even with the revision of the distance scale, though, cosmologists were very reluctant to envisage a value of the Hubble parameter much below 100 km/sec/Mpc. Determining this parameter is extremely difficult. It is easy to measure redshifts, but much harder to measure the distances to other galaxies, and thereby calibrate the redshift-distance relation. The only really reliable way is to study the behaviour of Cepheid variables in distant galaxies. Unfortunately, ground-based telescopes are not up to this task, and from the time of Hubble himself right into the 1990s the distances to other galaxies had to be inferred from a variety of secondary techniques, based on guesses about the brightnesses and sizes of galaxies. By the mid-1960s, a value for the Hubble parameter in the range 80–100, in the usual units, was widely accepted.

What baffled me about this, as I set out on the road to a career in astronomy, was that it still implied that the Milky Way is an unusually large spiral galaxy. Indeed, many astronomers (and textbooks) happily referred to the Milky Way as 'a giant spiral' (or words to that effect[2]), without seeming to consider the implications. But why should we just happen to live in the largest spiral galaxy around, when by that time it was clear that the Milky Way could not be so large as to be a continent among islands? If you guessed, indeed, that our Galaxy is an exactly average spiral, it would be possible to turn the whole argument on its head and work out the value of the Hubble parameter you would need in order to place all the other galaxies at the right distances for them to have an average size the same as the size of the Milky Way – remember, the smaller the value of the Hubble parameter, the further away other galaxies must be, and therefore the bigger they must be in order to look as large as they do in the sky. For the Milky Way to be an average spiral, the answer you get, from the proverbial back of the envelope calculation, is about 50 – if the value of the Hubble parameter were about 50 km/sec/Mpc, then all the other spirals would be ten times further away than Hubble thought in 1930, and ten times bigger than he inferred, so that their average size would be about the same as the size of the Milky Way.

This was obvious even to an MSc student in 1967, but there was no way, then, to test the idea. Over the next couple of decades, two schools of thought developed concerning the cosmic distance scale. One group of researchers, headed by Gerard de Vaucouleurs, favoured a value for the Hubble parameter of above 80, and a correspondingly short distance scale of the Universe, with (although this was not emphasized) correspondingly small galaxies. Another group, with Allan Sandage prominent among them, favoured a value of 60 or less, and a long distance scale of the Universe, implying larger sizes for the galaxies. It is some measure of how difficult this kind of work was that both teams used exactly the same observations, from which they drew different and incompatible conclusions.

Apart from the Principle of Terrestrial Mediocrity, the higher value of the Hubble parameter was always on more shaky ground (it seemed to me), because it ran into conflict with the ages of the oldest stars. Remember that the faster the Universe is expanding (the bigger the value of the Hubble parameter), the less time it will have taken to reach its present size, since the Big Bang. In the simplest cosmological models, a value for the Hubble parameter of 80 km/sec/Mpc would make the Universe only 9 or 10 billion years old; yet the oldest stars in our Galaxy were estimated by astrophysicists to be 14 or 15 billion years old. It was the same kind of dilemma that Hubble had faced when he found an age of the Universe less than the age of the Earth. But this was not a conclusive argument in favour of the long distance scale, because even a value of the Hubble parameter as low as 60 km/sec/Mpc still implied an age of the Universe of about 12 billion years, and this was still on the wrong side of the estimated ages of the oldest stars.

Perhaps because of inertia, having started out from an even higher value of the Hubble parameter and come down over the decades, it always seemed to me that most astronomers favoured the short distance scale and high value of the Hubble parameter, regardless of the difficulty with stellar ages; and whenever I tried to persuade anyone that the idea that we live in a typical galaxy would support the long distance scale and smaller value of the Hubble parameter, I wasn't taken seriously (except by people in the Sandage camp, but that was preaching to the converted). Of course, another reason I wasn't taken seriously was that in 1971 I stopped being a research astronomer and became a full-time science writer. In my capacity as a writer, though, in my book *Companion to the Cosmos* (published by Weidenfeld and Nicolson in 1996), I was careful to give the best estimate for the Hubble parameter as 55 km/sec/Mpc, largely on the (unstated) grounds that I believed the Milky Way to be an average spiral. This was something of an act of chutzpah, since just at the time that book was being written there had been a pronounced shift in favour of the short distance scale, thanks to early observations with the Hubble Space Telescope, which had been launched in 1990, but only became fully operational after the repair mission in December 1993. But things soon changed.

One of the principal aims of the HST, dignified by the name Hubble Key Project, was (and is) to measure the distances to other galaxies from the Cepheid technique, and thereby to get from their distances and redshifts a direct measure of the value of the Hubble parameter and the age of the Universe. To the consternation of those who, like myself, confidently expected the long distance scale to be proved correct, some of the earliest results from the HST, based on studies of a galaxy in the Virgo Cluster, suggested that the Hubble parameter must be large – above 80. We now know that these results were misleading, because the galaxy chosen for this study happens to have a large random velocity caused by the gravitational attraction of the other galaxies in the Virgo Cluster, so that its measured redshift is far from being purely cosmological. But it was in the wake of the debate spurred by the publication of these early HST results that, once again, I tried to persuade people of the importance of what was by then becoming known as the Principle of Terrestrial Mediocrity. And this time, someone listened.

By 1996, alongside my day job as a writer, I had been lucky enough to be appointed as a Visiting Fellow in Astronomy at the University of Sussex. This post, which I still hold, is purely an honorary position, with no pay and no responsibilities; but it provides me with some computer time, and contact with full-time researchers. When I started off once again on my hobby horse about the size of the Milky Way being a way to get a handle on the age of the Universe, two of those researchers, Simon Goodwin and Martin Hendry, took me seriously. They pointed out that it was no longer necessary to guess that the Milky Way is just an average spiral; there were already (this was in the summer of 1996) enough measurements of Cepheids in nearby spiral galaxies, from the Hubble Key Project to test the Principle of Terrestrial Mediocrity.

It took just one afternoon for Simon Goodwin to obtain the HST data over the Internet, giving us Cepheid distances of seventeen spiral galaxies. This number puts the whole problem in perspective – even with the HST, more than sixty years after Hubble's pioneering work, we had distances to just seventeen spirals from the Cepheid technique.

But it was enough. Knowing the distances to these galaxies, it was barely the work of a moment to compute their true linear diameters, and take the average. The result would have delighted Eddington. It turned out that the Milky Way, with a diameter of 27 kiloparsecs (just under 90 thousand light years) is slightly smaller than the average spiral galaxy – the average size for our sample of spirals was 28.3 kiloparsecs.

The next step was to look at a larger sample of spiral galaxies, ones which were too far away for even the HST to observe individual Cepheids in those galaxies, but which were still close enough to give big enough images in telescopes for their angular diameters to be measured. This stage of our investigation was delayed when Martin Hendry, the statistical expert on our team, moved to the University of Glasgow; but the wait was worthwhile. To our delight, we found that there were more than 3,000 spiral galaxies with known redshifts and angular diameters in an existing catalogue, compiled in 1991 and, like the HST data, available over the Internet. The distribution of the angular sizes of these galaxies (and therefore of their true linear sizes) follows a classic Gaussian distribution (the famous 'bell-shaped curve'), which proves that their diameters are randomly distributed around the average. And a battery of sophisticated statistical tests showed that the sample of nearby galaxies and the sample of distant galaxies both showed the same kind of Gaussian behaviour – that it was, therefore, meaningful to compare one with the other.

Knowing both the redshifts and the angular diameters of the galaxies on the sky, all we had to do to make the average of that bell-shaped curve for the distant sample match the average for our sample of nearby spirals with well-determined Cepheid distances was to choose a value for the Hubble parameter. The value needed to do the job is 52 km/sec/Mpc, with an error range (from the statistics) of plus or minus 6 km/sec/Mpc. In other words, the value of the Hubble parameter is between 46 and 58.

What is particularly delightful about this clearcut result is that it uses such a conceptually simple technique – exactly the standard technique used by a surveyor – but with data from one of the most sophisticated pieces of astronomical equipment yet built, the Hubble Space Telescope. Eddington saw the power of the argument in 1933;

I saw it in 1967. But nobody could actually apply the test until the HST had been in orbit long enough to provide Cepheid distances to the nearby spirals. We only needed seventeen of those distances – but there was no way we could have got them without the HST.

Using the simplest Einstein–de Sitter model of the Universe, with the critical density of matter required to make spacetime flat (omega equal to one) that value of the Hubble parameter would imply an age of the Universe of 12–13 billion years, since the Big Bang. This is the kind of model required by the simplest versions of the theory of inflation, which had looked to be in some difficulty in the light of other recent suggestions that the value of the Hubble parameter might be greater than 70. Our results are entirely compatible with simple inflationary models.

Even so, the age we derive on this basis would, just, still have been in conflict with the estimated ages of the oldest stars. But while we were carrying out this work, news came in of observations made by another orbiting observatory, Hipparcos. Hipparcos had been measuring the distances to stars using triangulation, with the advantage over Earth-based observations that the view it got was not blurred by the obscuring effect of the Earth's atmosphere. This was crucially important in pinning down the precise positions of stars to unprecedented accuracy. And Hipparcos found (the results were announced in 1997) that the stars on which these crucial age estimates are based are a little bit further away than used to be thought.

The way astronomers calculate the ages of stars is to compare their appearance today with the calculated appearance of a star the same size, based on computer models of how a star works. One of the key ingredients in the calculation is the brightness of a star – bigger stars burn more brightly and use up their fuel more quickly. The brightness of a star on the sky depends on its real, intrinsic brightness and its distances; the further away it is, the brighter it must shine in order to be visible on Earth at all. Because these stars are further away than used to be thought, they must be brighter, and bigger, than used to be thought. But bigger, brighter stars run through their life cycles more quickly. So, they would have needed less time to reach the stage of their life cycle they are in now, compared with the results of earlier calculations. The upshot of the Hipparcos studies is that the oldest stars in our Galaxy are now thought to be only 12 billion years old.

The two pieces of work mesh together beautifully. If the average size of all the nearby spiral galaxies (not just the Milky Way) is the same as the average size of all the spirals we can see, then the Universe is a minimum of 13 billion years old (any refinements to the cosmological models, other than the simplest Einstein–de Sitter model, increase the estimated age). And the oldest stars are 12 billion years old, so there was a billion years or so after the Big Bang in which the first stars could form. The Big Bang really did happen.

Notes

Chapter 1 *The Arrow of Time*

1 There is also a 'zeroth' law of thermodynamics, an afterthought put in by a later generation of scientists, which concerns the definition of temperature; for my purpose here, we can manage quite happily with the everyday understanding of temperature as a measure of hotness. The third law of thermodynamics has to do with the impossibility of matter ever cooling to the ultimate low temperature, the absolute zero (just under −273 °C). Again, these subtleties are not important here. Of the four laws of thermodynamics, the second is the one that is important to an understanding of the evolution of the Universe.

2 To the mathematical physicists, it is information, or order, that is described in terms of negative entropy, or as 'negentropy'.

3 There is a convention by which Universe, capitalized, refers to the real world of stars and galaxies that we can see, while universe, without the capital letter, refers to one of the more or less speculative ideas of the theorists. I shall try to follow this convention. Similarly 'Galaxy' means our Milky Way, while 'galaxy' refers to any old galaxy anywhere in the Universe.

4 and 5 Both quotes are taken from page 254 of *Order out of Chaos*, by Ilya Prigogine and Isabelle Stengers. They originate in Boltzmann's *Lectures on Gas Theory*, reprinted by the University of California Press in 1964.

6 If you want the full story of the quantum revolution, see my book *In Search of Schrödinger's Cat*.

7 Even atoms are relatively large by the standards of quantum physics, so the effect of quantum uncertainty is still small at this level. But the same argument applies to the deeper level of electrons and protons, so the argument is completely valid. And, of course, even the tiniest deviation from perfection is still imperfection.

Chapter 2   *The Realm of the Nebulae*

1 *The Realm of the Nebulae*, pages 23 and 24. Full details of books mentioned in the text are given in the bibliography.

Chapter 3   *How Far Is Up?*

1 If one observer views the Moon directly overhead, and the other sees it low on the horizon, at the same time, the parallax shift between their two lines of sight is 57 minutes of arc, nearly twice the Moon's angular diameter. The baseline for the triangulation is then equal to the radius of the Earth, and this gives the distance to the Moon.

2 Red, orange, yellow, green, blue, indigo and violet, in order of decreasing wavelength.

3 The stars in such a cluster appear packed together when viewed through a telescope, and they are close to one another compared with stars that are not in clusters, but there is still plenty of space between them. In the dense heart of a globular cluster, stars are separated from one another by, on average, about one-tenth of the separation between stars in the neighbourhood of the Sun.

4 Hardly surprisingly, there was for many years an intense, sometimes bitter, rivalry between the two great observatories in California, Lick associated with the University of California and Mount Wilson associated with the Carnegie Institution and later with Caltech. The hatchet has, however, long since been buried.

5 Ironically, Curtis, while right about the distance to the Andromeda galaxy, was wrong on almost every point of his view about our own Galaxy. He thought the Sun was at the centre of the Milky Way, and made an estimate for the size of the whole system which was much too small. To get the best picture of the Universe in 1920, you needed half of Shapley's picture and half of Curtis's.

6 Along the way, the spinoff from the project had a major impact on American, and world, astronomy. The Corning Glass Company of New York, commissioned to create the 200-inch mirror, decided to learn the tricks of the trade by casting a series of smaller mirrors first. These test castings were not wasted – two 61-inch mirrors went to Shapley at Harvard; a 76-inch mirror went to Toronto; an 82-inch to the McDonald Observatory in Texas; a 98-inch to the observatory at the University of Michigan; and a 120-inch to Lick.

Chapter 4   *The Expanding Universe*

1 This is a continuing story, with new breakthroughs in the 1980s giving astrophotographers the best views yet of astronomical objects, and the best spectra from them. David Malin and Paul Murdin describe many of these techniques, and present some breathtaking photographs produced by them, in their book *Colours of the Stars*, published by Cambridge University Press in 1984.

2 Quoted, with no source given, by Denis Sciama in *Modern Cosmology*, page 43.

3 There are good reasons for this, of course, and the caricature I sketch here doesn't really do justice to the theorists. They were right to feel tentative about the early, simple cosmological models, which hardly looked sophisticated enough to account for the complexity of the observed Universe. It was only gradually, over decades, that reasons began to appear for these almost childish models to be recognized as a reasonable indication of the nature of the Universe. The reasons for believing in those simple models have now become highly sophisticated; but it remains a great and fundamental truth that the Universe seems to run according to rules that are so simple that the theorists initially felt that they *must* be inadequate descriptions of reality.

4 See, for example, *Albert Einstein*, by Banesh Hoffman, page 32.

5 See my book *In Search of Schrödinger's Cat* for the story of the roots of quantum physics. I deal in more detail with some of the non-cosmological implications of relativity theory in my book *In Search of the Edge of Time*.

6 And light from the Andromeda galaxy, which has a blueshift corresponding to a large velocity of that galaxy towards us, still only has a velocity c relative to us.

7 Even if there are intelligent beings living on a planet in one of those galaxies receding from us at one-third the speed of light, when they measure the speed of the light leaving their galaxy it is c, and when we measure it as it arrives here it is still c, not $\frac{2}{3}c$. The moral of all this is still relevant to scientific research today. As Bill McCrea pointed out when commenting on this chapter, Newton's ideas must have seemed exotic and counter to 'common sense' to people who had never met mathematical physics before. Newton invented mathematical physics, which was a revolutionary innovation at the time. Einstein elaborated on the basic ideas Newton put forward, but he used the same techniques that Newton had invented. So Newton was, perhaps, the greater innovator. And the moral? It is not to regard *anything* as 'common sense' or 'obvious', but to approach everything with an open mind and without preconceptions – that is what Einstein did, and that is the genius behind Einstein's work.

8 It doesn't even have to be a *constant* force; but let's keep it simple for now.

9 It is important to distinguish the effects of velocity and of acceleration here. If the mythical space lab is floating with a steady velocity, then anything crossing the lab from one side to the other has the same velocity, and although both the lab and the object move forward, they move by the same amount. It is acceleration, a changing velocity, that makes the lab move an extra amount forward while the light beam is travelling across it. In the falling elevator, the light beam is also under the influence of gravity and is accelerated, so the appropriate thought experiment really is a lab in space being pushed by some outside force, like its rockets, not one falling under the influence of gravity.

10 McCrea says he doubts whether Einstein actually derived his ideas *starting* from the principle of equivalence. He suggests that this is simply a useful device Einstein used in expounding his ideas after he had developed them. But McCrea also says that most of his colleagues would disagree with this interpretation of the evidence, and so I have stuck by the traditional version of the story here.

11 In, of course, the *Annalen der Physik* (Volume 49, page 769). The paper is called 'The foundation of the General Theory of Relativity', and is widely accepted as the greatest paper Einstein ever wrote.

12 There are several versions of this story that give slightly different emphases but report the same series of events. I have followed the account given by S. Chandrasekhar in his book *Eddington*, which is also the source of the direct quotes from and about Eddington given here.

13 Following the two expeditions of 1919, the next eclipse study was in 1922, when a Lick Observatory expedition to Wallal, in Western Australia, provided the third measurement of the Einsteinian light bending effect, and again showed the results predicted by General Relativity. The news was announced to the April 1923 meeting of the Royal Astronomical Society, where Eddington, characteristically, quoted Lewis Carroll: 'I think that it was Bellman in "The Hunting of the Snark" who laid down the rule "When I say it three times, it is right." The stars have now said it three times to three separate expeditions; and I am convinced their answer is right.'

14 McCrea gives a very good, readable account of General Relativity in his contribution to the volume *Cosmology Today*, which I edited. I would not have ventured to describe General Relativity here without checking this chapter with McCrea, who says that my account is 'quite defensible'. I *think* that means it is OK, if not quite the way he would have told the story himself!

15 Quotes from McCrea in *Cosmology Today*.

16 The 'Newtonian cosmology' invented by Milne and McCrea cannot really cope with the problem of edges, because it doesn't allow for bent spacetime, so it doesn't really provide a valid description of the Universe – or any universe. So it cannot be said in any real sense to be a rival to relativistic cosmology. But it does have its uses.

Chapter 5   *The Cosmic Egg*

1 Quoted by North, page 73. In a posthumously published work, *Common Sense of the Exact Sciences*, Clifford went even further, speculating that space may be curved evenly in all directions but that 'its degree of curvature may change as a whole with time'. This is a remarkable precursor of the expanding universe of relativistic cosmology. See Edward Harrison, *Cosmology*, page 155.

2 The aphorism is widely quoted, and generally attributed to Eddington. Although it is just possible that he picked it up from someone else, it has his characteristic ring about it, and it certainly appeared in a paper of his in the *Proceedings of the Physical Society* in 1932 (Volume 44, page 6).

3 They were published in the German journal *Zeitschrift für Physik* in two papers in 1922 and 1924.

4 The Einstein–de Sitter model is a special, very simple, solution to the Friedman equations.

5 See Gamow's autobiography, *My World Line* (New York: Viking Press, 1970).

6 Lemaître's original paper appeared in the *Annals* of the Scientific Society of Brussels (Volume 47A, page 49). Eddington's enthusiastic espousal of the work led first to a review of this paper by Eddington himself, which appeared in an article in the *Monthly Notices* in 1930 (Volume 90, page 668), and then to the English translation of the original, which appeared in 1931 (Volume 91, page 483).

7 P. J. E. Peebles, *Modern Cosmology* (Princeton, NJ: Princeton University Press, 1971), page 8.

8 It is better to think in terms of the separation of any two distant galaxies, rather than in terms of the recession of distant galaxies 'from us'. Our measurements of redshift have to be made from here on Earth, and do show up as a recession of distant galaxies from us, but this is misleading. We are not at some special place, the centre of repulsion in the Universe. It is simply that the entire fabric of spacetime is expanding, so that *all* observers, living anywhere in the Universe, see the same effect. Think of a perfectly smooth bubble, marked with tiny spots of paint. As the bubble expands, *every* spot gets further away from every other spot, and the 'view' from one spot – any spot – will be of all the other spots receding from it. In his 1917 paper on cosmology, Einstein proposed a fundamental postulate, that there is no average property of the Universe which defines either a special place in the Universe or a special direction in the Universe. This is known as the Cosmological Principle, and in effect it says that we live in a typical, ordinary region of the Universe and that the view we get is just the same, on average, as the view anyone else would get, anywhere else in the Universe. Of course, it may be that Einstein started out from this fundamental assumption for no more subtle reason than that it provided the simplest model he could think of, and he wanted to try his equations out on simple models, at least to start with. It is the growing realization that our Universe seems to fit this simplest possible description that has elevated the Cosmological Principle to its present exalted status.

Uniform expansion of spacetime (or uniform contraction, which is its mirror image mathematically) is the only kind of dynamic distortion of the Universe that conforms to the Cosmological Principle – an impressive indication that Einstein's postulate (or guess) is telling us something fundamental about our Universe, especially bearing in mind that when he laid down the Principle he did not know the Universe was expanding. Had his thoughts turned in that direction, he could have predicted, from the Cosmological Principle alone, that if astronomers were to discover large-scale motion in the Universe then they would discover either uniform expansion, or uniform contraction, and nothing else. Indeed, the equations *did* make this prediction – until Einstein stopped them by putting in his cosmological constant.

Subsequent observations of the distribution and large-scale motion of matter in the Universe provide strong circumstantial evidence in favour of the Cosmological Principle, but it is one of those things that can never actually be *proved* correct. However, if the Universe were constructed in such a way that it looked very different to different observers, there would not be much point in doing cosmology at all, since we would not be able to deduce anything about the Universe at large from observations we make in our little corner of it. Without the Cosmological Principle, in a very real sense, there would be no cosmology.

9 The best, and philosophically most aesthetically pleasing, proposed resolution of the time scale difficulty came in the late 1940s, from Hermann Bondi, Tommy Gold and Fred Hoyle. They came up with the idea of an *expanding* Steady State universe, in which new matter was constantly being created to fill in the gaps left between the old galaxies as time passed and the universe grew. The wonderful simplicity of this idea appealed to many mathematician cosmologists, while many observational astronomers realized that the Steady State theory at least provided them with a target to shoot at, whether or not they liked the philosophy behind it, and whether or not they cared if it was 'right' or 'wrong'. The only definite cosmological test they could carry out was to see if any observation would invalidate the Steady State model (it is much harder to invalidate the Big Bang idea, of course, since almost anything, including a long interval of Steady State behaviour, fits in with one of the Big Bang models!). So the Steady State idea stimulated research over two decades. It gave the observers predictions that could be tested, and suggested observations that would decide between the Steady State and Big Bang models. By stimulating observers to redoubled efforts, the Steady State model undoubtedly hastened the pace of cosmological research. Unfortunately for those, like myself, who still find the philosophy behind it attractive, it fell by the wayside when increasing evidence that there really was a Big Bang came in. But though it may have been an *incorrect* hypothesis, it was certainly a *good* hypothesis – one which could be tested and which hastened the development of scientific understanding.

10 There's a nice irony in the fact that Einstein rejected his cosmological constant largely as a result of Lemaître's work, while Lemaître retained the cosmological constant to get what he regarded as a satisfactory model!

11 Since the 1950s, the Universe has been probed not just with optical telescopes but by radio telescopes as well, and lately by X-ray and other instruments carried above the Earth's atmosphere. Radio observations played an important part in determining that the Universe is evolving, but the detailed history of radio astronomy is outside the scope of this book.

12 All this talk of spherical and hyperbolic spaces may make you wonder whether there are other alternatives, comparable, perhaps, to the ellipses. The analogy doesn't really follow through at this level, and it is best at this stage to concentrate on the distinction between hyperbolic (open) models and spherical (closed) models. But cosmologists do sometimes dabble with the equations of a space with a slightly different topology, which they call elliptical space. This has several interesting properties. It has only half the volume of the equivalent spherical space with the same 'radius'. This is because although in spherical space every point has its equivalent antipodal point (like the North and South Poles of the Earth), in elliptical space when you travel to the furthest 'point' from a chosen point you could be anywhere in a region of space, in the same way that all points on the equator are at the same distance from the North Pole. But even if the spacetime we live in is elliptical in this sense (and there is no evidence that it is) that doesn't affect the puzzle of why it is expanding and how it was born in a Big Bang.

13 Just what these differences, and the fine tuning, mean for the fate of our

Universe is an important topic which is outside the scope of this book. But it is perhaps worth mentioning here that a universe which is closed by gravity is precisely the equivalent of a black hole. Indeed, it *is* a black hole – or, putting it another way, the black holes that caught the public imagination so much in the 1970s and 1980s are pocked universes, described completely by Einstein's equations for a region of spacetime closed by gravity.

14 A fuller account of all this pioneering work in particle physics is provided in my book *In Search of Schrödinger's Cat*. The sketch given here is scarcely more than a caricature, but one which I hope shows how physics and cosmology developed alongside one another.

15 And a fuller account of how chemistry 'works' at the atomic level can be found in my book *In Search of the Double Helix*.

16 And just as you *can* kick the ball higher up the slope, if you put enough energy into the kick, so it is *possible* to make very heavy elements by pushing energy into the fusion process. Equally, just as the ball will sit happily halfway up the slope if left alone, so many nuclei are quite stable, even though they are, strictly speaking, in a more energetic state than the iron-56 minimum. All this is important to an understanding of how the elements formed in the first place, a main theme of Chapter 6.

17 The half life is the time it takes for half of the particles in a sample to decay. If you start with a hundred neutrons, after 13 minutes there will be fifty left, after 26 minutes there will be twenty-five left, after 39 minutes a dozen, and so on. This kind of statistical behaviour, in which the fate of an individual particle cannot be predicted (there is no way to tell *which* of the original neutrons will decay in the first 13 minutes, or any subsequent period) but the overall behaviour of a large group (it should really, of course, be far more than a hundred) is completely predictable, is one of the most striking aspects of quantum physics. It applies to all radioactive decays, and to other processes, not just to the decay of the neutron.

18 Where did the dense soup come from? In the 1940s (and, indeed, in the 1950s, 1960s and 1970s) that question was unanswerable. Gamow got round the problem by invoking one of the Friedman models which starts out very thin, collapses down into the cosmic soup stage, and then expands again. But there are now other resolutions of this dilemma.

19 The image of a smoothly sloping valley with iron-56 at the bottom is, of course, an oversimplification. The helium-4 nucleus, to extend the analogy, sits in its own hole in the ground on the side of the valley, from which it cannot easily be dislodged; the nucleus with eight particles might be thought of as balanced on top of a molehill on the valley side (a very sharply pointed molehill!) from which it topples at the slightest provocation.

20 *The Creation of the Universe*, page 65.

21 The zero of the Kelvin scale of temperature is −273 °C, to the nearest whole number. It is the absolute zero of temperature at which all particles are in their lowest energy levels, called zero-point states; there can be nothing colder.

22 Andre Deutsch edition, pages 131 and 132.

Chapter 6   *Two Keys to the Universe*

1 One way to visualize this is to extend the spring analogy. If you wanted to spread the Sun back out into a thin cloud, you would have to do work, against the pull of gravity, on every particle in it to move it further away from the centre of mass. The particles in the diffuse state store up that energy as gravitational potential energy. And that is the energy liberated as heat when such a cloud contracts.

2 One implication of the Kelvin–Helmholtz effect is that it would take about 10 million years for the Sun to adjust to any drastic change in the heating processes in its interior. If those processes, *whatever* they may be, were suddenly switched off, we here on Earth wouldn't notice for several million years, because gradual gravitational collapse would keep the Sun hot. A comforting thought, perhaps, in these uncertain times.

3 Inside a star like the Sun, the raw material is all hydrogen nuclei, protons. Half of these have to be turned into neutrons along the way, which involves ejecting a positron to carry off the 'spare' electric charge. Positrons are the positively charged counterparts to electrons, and when a positron meets an electron the pair annihilate and all their mass is converted into energy. The result is that the numbers that go into the calculation are slightly different, and the calculations are that little bit more complicated, but you still end up with the 'answer' that just over 0.7 per cent of the original mass is turned into energy every time a helium nucleus is manufactured.

4 At the same time, in 1938, another German physicist, Carl von Weizsäcker, proposed the same basic mechanism for keeping the Sun hot. Like Bethe, von Weizsäcker also worked on the problem of making an atomic bomb during the 1940s – but he was working in Germany at the time, and this may explain why he did not share the Nobel award with Bethe in 1967, even though it has been suggested that von Weizsäcker did everything in his power to ensure that the work of his team did not, in fact, provide the Nazis with nuclear weapons.

5 To be strictly accurate, only the CN part of the cycle was proposed in 1938–9; the minor role played by oxygen, and the change of name to CNO cycle, came later. There are really two cycles, involving CN and NO, respectively.

Some astronomers, with a shameless addiction to puns, therefore refer to the CNO 'bi-cycle'.

6 The address is reprinted in *Science*, Volume 226, page 922.

7 *Monthly Notices of the Royal Astronomical Society*, Volume 106, page 343.

8 Volume 29, page 547. Alastair Cameron, a Canadian-born American astrophysicist, came to similar conclusions independently and published his calculations in 1957 in the *Publications of the Astronomical Society of the Pacific*, Volume 69, page 201. If it seems a little unfair that he should be relegated to a footnote, remember that it was, after all, Hoyle who came up with what Fowler calls 'the grand concept of element synthesis in the stars' (Nobel address, *op. cit*), and it was Fowler's team who carried through the reaction rate studies in the lab. History probably is correct to give B^2FH pride of place, though you might think the Burbidges were lucky to have been in the right place at the right time to join Fowler and Hoyle in their endeavours.

9 The two were, scientifically, almost inseparable. Fowler, who died in 1995, published between 200 and 300 scientific papers in his career; of these, no less than 25 (about 10 per cent) carry Hoyle's name as well. It is, indeed, one of the worst blunders ever made by the Nobel Committee that they did not see fit to keep the two names together in 1983. And, with all due respect to Fowler, if they had some reason for only mentioning one member of the team, then without doubt that should have been Hoyle.

10 And being examined for my PhD, to my own amazement, by an unusually distinguished pair of interrogators, two of the great scientists who enter into this story, Fowler and McCrea. The rules at Cambridge do not allow the oral examiners to tell the student if this crucial final interview has been a success, and news of the outcome is supposed to await official written confirmation from the authorities. But the examiners usually find a way to reassure the nervous student. On this occasion, a hot summer's day, McCrea, Fowler and myself emerged from the small, sticky office where I had been trying desperately to recall what little I remembered about the work I had been doing for the past three years, and headed off to join the gathering throng at tea. Fowler had his tie askew, top shirt button undone, and jacket slung over his shoulder. As he mopped sweat from his brow, a colleague, Stirling Colegate, jovially enquired, 'Well, Willy, did he win?' Back came the response, 'I'm not allowed to tell you.' 'Well,' said Colegate, in a slow drawl, 'it sure looks like you lost!'

11 *Physics Bulletin*, Volume 35, page 17. McCrea surely knows what he is talking about. Like Hoyle, he was one of the pioneers in developing the Steady State model, and as a good scientist who thinks deeply about the nature of the scientific endeavour he still cautions that the Big Bang model is not 'proved' correct. Simple Big Bang and simple Steady State models represent two extremes in a vast range of possibilities, and there are many shades of grey in between. The important thing, he says, is to discuss all the possibilities with an open mind – and then, although one model may seem a better description than any other of the world we live in today, it is unlikely that any scientist will be able to say, with absolute conviction, 'this is right'. But the hot Big Bang model *is* accepted as by far the best model of the Universe by the great majority of astronomers today. The irony that Hoyle, chief proponent of the Steady State model, should be co-author of a landmark paper which persuaded people of the validity of the Big Bang model gives a wry twist to the tale.

12 G. Herzberg, *Spectra of Diatomic Molecules*, second edition, Van Nostrand, Princeton, 1950.

13 And since reprinted in the collection *Observing the Universe*, edited by Nigel Henbest (Blackwell, Oxford, 1984). The quote is taken from page 9 of that volume.

14 Maser is an acronym for microwave amplification by stimulated emission of radiation; the important words in the present context are 'microwave' (which means it operates at wavelengths of a few centimetres) and 'amplification' (which means it makes a weak radio input stronger). A laser is the same kind of thing, but operating with light instead of microwaves.

15 But it is *possible* to provide alternative explanations for the background radiation. David Layzer, of Harvard University, has recently argued vociferously that it may be produced by radiation from a generation of massive stars that formed *before* galaxies formed, in an initially cold universe. Such possibilities are certainly worthy of discussion, and serve as reminders that the accepted 'best buy' model of the Universe does not necessarily represent the last word. Layzer makes the case for his version of cosmology in his book *Constructing the Universe*; unfortunately (if understandably) he fails to provide a balanced view, and presents his version as the 'best buy', while dismissing the standard model almost out of hand.

16 *The First Three Minutes*, Deutsch edition, page 132.

Chapter 7 *The Standard Model*

1 Mathematicians will tell you that we cannot literally start our clock at $t = 0$, because there is no way of getting out of the singularity, and there is no 'instant' $t = 0$. The concept of time has no meaning at the singularity itself. My answer to the mathematicians is that we can set $t = 0$ as close as we like to the singularity. For the present discussion, the difference is not important. It *is* important when we bring in the effects of quantum physics and try to deal with the moment of creation itself, and the results of this kind of investigation are described later in this book.

2 I. D. Novikov's book *Evolution of the Universe* gives a slightly more detailed and technical, but still very readable, account of the first instants, written in 1978. If you read this and Weinberg's book together, you will get a very clear picture of the understanding of the Universe cosmologists had at the end of the 1970s.

3 So you need a photon with an energy of more than 1 MeV in order to create an electron-positron pair.

4 Except through gravity. The energy stored by the neutrino sea itself contributes to the gravity of the whole Universe, and there have been speculations that neutrinos may not be completely massless, but might each carry a mass of a few eV. If so, because there are so many of them the total mass would be a major fraction of the total mass of the Universe, with profound implications for the ultimate fate of the Universe, perhaps ensuring that it is closed in spite of the implications of the deuterium abundance for the density of nucleonic matter.

5 Heavier nuclei do form early on, at high densities, but are broken apart again and never 'freeze out' of the fireball. Only hydrogen, helium, a little deuterium and a trace of lithium, plus neutrinos and other nonnucleonic particles emerge from the Big Bang.

6 The redshift z is defined in terms of the amount by which a spectrum has been stretched. If a feature in the spectrum occurs at a certain wavelength, λ, as measured by someone who is stationary compared with the source of the spectrum, then an observer who is moving in such a way that the source of the spectrum seems to be receding from him will measure the same feature at a longer wavelength. The difference between the two wavelengths is written as $\Delta\lambda$; and $z = \Delta\lambda/\lambda$. The yellow sodium D line, for example, occurs at a wavelength of 589 nanometres in the laboratory. If we examine the spectrum of a star, or any other object, and find that the sodium D line is at a wavelength of, say, 600 nanometers, then the redshift of that object is 11/589, that is, $z = 0.01868$. For such a small redshift, the implied velocity of recession is simply the speed of light, c, times z, which comes out as about 5,000 kilometres per second, for this hypothetical example. For redshifts bigger than about 0.4, it is essential to use a slightly more complicated formula, taking account of relativistic effects. This says that for an object receding at velocity v:

$$(1 + z) = \sqrt{(c + v)/(c - v)}$$

At a redshift of 2, this formula gives the correct recession velocity of $0.8c$, 80 per cent of the speed of light. But no matter how big the redshift is, v can never exceed c.

Chapter 8 *Close to Critical*

1 *More* energy escapes from the red giant, even though its surface is cooler than our Sun's surface, because the surface area is so much bigger. The amount of energy crossing each square metre is less, but there are many, many more square metres each contributing their share. A red giant actually emits a hundred times as much energy as our Sun does today.

2 Curiously, in July 1991, with interest in the background radiation stimulated by the COBE discoveries, another twist in the tortuous saga of the history of astronomers' failure to notice the cosmic background radiation surfaced in the pages of *Nature* (Volume 352, page 198). It seems that as long ago as 1955 a French research student, E. Le Roux, measured the 'temperature of the sky' as 3 ± 2 K, using war-surplus radar equipment, for his PhD studies. He suggested that the origin of the radio noise he had measured might lie beyond our Galaxy, but he did not make the connection with the predictions by Gamow and his colleagues of a cosmological background, and his findings appeared only in the pages of his thesis, on the shelves of the library of the University of Paris. Yet another near-miss on the Nobel Prize!

Chapter 9 *The Need for Inflation*

1 Because they combine to form atomic nuclei, protons and neutrons are together also known as nucleons. They are members of a larger family of particles known as baryons – the baryons from which atomic nuclei are made. There are other kinds of baryons, which can be manufactured in colliding-beam experiments at particle accelerators, but they do not occur in significant numbers in the Universe at large today. For the purposes of the present discussion, the terms 'nucleon' and 'baryon' are interchangeable.

2 The figure comes from the work of David Schramm, of the University of Chicago, who presented it to a meeting of the Royal Society, in London, in October 1982.

3 The significance of this epoch, 10^{-35} second after the moment of creation, will be explained in Chapter 10.

4 Quote from Guth's contribution to *The Very Early Universe*, edited by G. W. Gibbons, S. W. Hawking and S. T. C. Siklos, page 201.

5 *Physical Review*, Volume D23, page 347, published in January 1981.

6 Just as 10^2 is 100 (10×10), so 2^2 is 4 (2×2) and so on; 2^{100} means 2 multiplied by 2 one hundred times. Another way of writing this would be as twice $(2^3)^{33}$, and 2^3 is 8, so that would be 8^{33}, or 8 multiplied by 8 thirty-three times. If we fudge round the edges of the numbers a little, to get a rough handle on what exponential inflation means, we can say that 8 is near enough 10, and that 2^{100} is near enough 10^{33}. In the same way, inflation by a factor of 10^{50} corresponds to doubling roughly 150 times – that is, growth by a factor of 2^{150}. And with one doubling every 10^{-34} second, that takes only just over 1.5×10^{-32} second (150×10^{-34} second) to complete.

7 A light year is 9.5×10^{27} cm, but the expansion of the Universe can stretch spacetime much 'faster than light' because nothing is moving *through* spacetime. So the exponential inflation of the Universe proposed by Guth and other cosmologists really can take a region of spacetime much, much smaller than a proton and blow it up to a volume 100 million light years across in a tiny fraction of a second.

Chapter 10   *In Search of the Missing Mass*

1 These particles are actually *manufactured* out of energy in machines that accelerate everyday particles, such as protons and electrons, to very high energies and smash them into one another. The more massive families of particles would exist naturally in a high-energy world (or early in the Big Bang), but they do not occur naturally in the Universe at large today, and when they are made artificially they soon decay into more familiar particles of the everyday world.

2 Evidence for the 'strange' quark had actually come in the 1960s. But it was only in the 1970s that physicists appreciated the need for 'charm' as well, and realized that nature was duplicating the basic quark/lepton theme.

3 Perhaps I should say 'very little' chance. Some physicists have speculated that a form of matter containing equal numbers of up, down and strange quarks might be stable. If so, and *if* such matter was produced in the Big Bang before the epoch of baryon formation, then there *might* be enough of this 'strange matter' about today, perhaps in the form of 'quark nuggets', to play a part in providing the dark matter. But I wouldn't bet on it.

4 *Exactly* the speed of light, if their mass is precisely zero.

5 Why the delay, compared with Lee and Yang? Well, by the mid-1960s, the discovery of a violation of one of the conservation 'rules' wasn't such a big shock as it had been in 1956!

6 Sakharov's name is familiar to a Western audience largely for political reasons, because of his position as a prominent 'dissident' in the 1970s. But this publicity for his political activities as a campaigner for nuclear disarmament and human rights has, if anything, obscured the fact that he

was one of the most brilliant physicists of his generation. Born in Moscow in 1921, Sakharov followed in his father's footsteps to become a physicist, graduating from Moscow State University in 1942 and joining the P. N. Lebedev Institute in Moscow in 1945. He was closely involved with the development of thermonuclear weapons in the Soviet Union, and in 1953 he became the youngest person ever, at the age of thirty-two, to be made a full member of the Soviet Academy of Sciences. He became a Hero of Socialist Labour, and was awarded the Order of Lenin and many other honours. But by the late 1950s he was becoming interested in social issues.

At first, his public comments were devoted to ideas for the reform of the Soviet educational system, which, while not following the party line, were published in *Pravda*, and some of which became official policy in the early 1960s. During the 1960s, Sakharov's scientific research concentrated on cosmology and the initial stages of the Big Bang. In 1968, he published an essay calling for a reduction in nuclear arms, and in 1970 he was one of the founders of a Committee for Human Rights. His increasing political activity during the 1970s led, among other things, to the award of the Nobel Peace Prize in 1975, and to banishment to Gorky in 1980. It looked at that time as if his scientific career was at an end. But in 1984, following the development of new cosmological ideas which echoed some of Sakharov's theoretical work in the 1960s, papers from Sakharov taking up these ideas and discussing fundamental concepts related to the creation and early evolution of the Universe appeared in the Soviet *Journal of Experimental and Theoretical Physics*.

7 Sakharov's insight, explaining the requirements that had to be met in order for matter to exist in the Universe today, and made without the aid of the GUTs that were to come in the 1970s, surely deserved a Nobel Prize, but alas, such awards are never made posthumously, and Sakharov died in 1989.

8 When astronomers study the light from very distant quasars, they find that the bright spectrum of the quasar is crossed by many dark lines corresponding to many 'copies' of the spectrum of cold hydrogen, but shifted by different amounts to the red. This is called the Lyman forest; it is interpreted as being caused by light from the quasar being absorbed in many different clouds of cold hydrogen at different distances (different redshifts) between us and the quasar. These clouds could be 'failed' galaxies, masses of hydrogen trapped by WIMP potholes but which never formed bright stars. If that interpretation is correct, the voids between the froth of bright galaxies may indeed be full of cold, dark matter.

9 Quantum physics, indeed, says that even the location of the mathematical point cannot be precisely known.

Chapter 11 *The Birth and Death of the Universe*

1 If our descendants survive this catastrophe and leave the Earth and Sun behind, they will by then surely no longer be 'human' in the way that we are!

2 Borrowed, in part, from the excellent (but weighty) book *The Anthropic Cosmological Principle*, by John Barrow and Frank Tipler.

3 Assuming, for the sake of argument, a proton 'half-life' of *exactly* 10^{31} years. This is less than the limits set by the latest experiments, but the argument is the same whatever the exact numbers you put in.

4 *Nature*, Volume 246, page 396; this paper was actually stimulated by an article I had written. A similar idea was put forward independently in the Soviet Union at about the same time, by P. I. Fomin, and circulated as a 'preprint'; but this work was not published until 1975.

5 The basic idea of something for nothing goes back at least thirty years before Tryon's version appeared. In his autobiography *My World Line*, George Gamow tells about his role during World War Two as a consultant with the Bureau of Ordnance in the US Navy Department in Washington, DC. Gamow was not allowed to work on the atomic bomb, because he was a Russian by birth and delighted in telling all his friends that he had been a colonel in the Red Army at the age of twenty. Even allowing for Gamow's predisposition to tall tales and practical jokes, this wasn't to be taken lightly by those responsible for security on the Manhattan Project. So he spent the war years in Washington.

One of Gamow's jobs, however, was to deliver a briefcase full of papers to Albert Einstein, in Princeton, once a fortnight. Although officially secret, these papers had nothing to do with nuclear weapons. They described all kinds of ideas for new weaponry, which Einstein was expected to comment on for the Navy. Einstein would almost invariably comment favourably, as Gamow tells it, no matter how weird and wonderful the new ideas for explosive devices. One day, while walking with Einstein from his home to the Institute for Advanced Study, Gamow mentioned that Pascual Jordan had come up with a new idea. Jordan was one of the pioneers of quantum physics, who made his name through work with Werner Heisenberg and Max Born which established the basis for the first version of quantum mechanics, called matrix mechanics, in 1925. His new idea didn't seem, in the 1940s, to be in that league. It was, indeed, just one of those crazy ideas physicists like to mull over during coffee time, or while walking through Princeton. What Gamow mentioned to Einstein on their walk was Jordan's idea that a star might be created out of nothing, since at the point zero its negative gravitational energy is numerically equal to its positive rest mass energy.

'Einstein stopped in his tracks,' Gamow tells us, 'and, since we were crossing a street, several cars had to stop to avoid running us down.'

The idea that stopped Einstein in his tracks is the same idea, now applied to the whole Universe, not just a star, that researchers such as Ed Tryon are taking very seriously indeed in the 1990s.

6 Vilenkin's unusual career development is worth a brief mention. He was born in Kharkov, in what was then the USSR, in 1949, and obtained his BSc in 1971, from Kharkov State University. But, he told me, he was unable to obtain a research post because he is Jewish, and he spent five years first in army service and then earning a living at various odd jobs (his favourite, he says, being night watchman in a zoo) before emigrating to the US in 1976. During those five years, however, he had been studying physics in his spare time, to such good effect that in 1977, the year after he arrived in the States, his work earned him a PhD from the State University of New York, Buffalo.

7 Not just his work on relativity theory, either. The first time I visited Cambridge was in 1967 to attend a presentation of the then-new ideas on primordial nucleosynthesis, from the Wagoner, Fowler and Hoyle team. In a packed lecture room filled with some of the best physicists and astronomers in England, where the role of a research student such as myself was to keep quiet and take notes, the most penetrating questions from the audience were asked by a young man I had never seen before, a junior researcher who seemed to have a slight speech impediment but clearly had a first-class grasp of the subject under discussion. He was, of course, Stephen Hawking.

8 This award is generally regarded by physicists as the ultimate accolade, ranking significantly above the Nobel Prize.

9 It is only right to point out that few physicists are entirely happy with Hawking's approach. He has to make a lot of simplifying assumptions, and they don't always approve of the way he handles the equations. But the underlying physical principles of the model are very clear and straightforward, and it is this that persuades me that Hawking is on the right track. The details of the equations may change; I doubt if the underlying physical principles will.

10 The quote comes from a paper by Tipler in *Physics Reports*; he covers the same ground in part of Chapter 7 of his book with John Barrow.

11 In John Paul II's own words, addressing that meeting in 1981, 'any scientific hypothesis on the origin of the world, such as that of a primeval atom from which the whole of the physical world derived, leaves open the problem concerning the beginning of the Universe. Science cannot by itself resolve such a question: what is needed is that human knowledge that rises above physics and astrophysics and which is called metaphysics; it needs above all the knowledge that comes from the revelation of God.' And he went on to

quote a predecessor, Pope Pius XII, who said in 1951, referring to the problem of the origin of the Universe, 'We would wait in vain for a reply from the natural sciences, who on the contrary admit that they are honestly faced with an insoluble enigma.' Forty years later, that 'vain wait' is over. Many cosmologists no longer accept that the enigma is insoluble and there is a feeling now that science *can* resolve the metaphysical puzzle of the origin of the Universe. Hawking's universe, ironically presented at that very gathering where John Paul II claimed a still pre-eminent role for God, clearly points the way to an ultimate *scientific* solution to the greatest metaphysical puzzle of them all.

Appendix　*The End of the Search*

1　Arthur Eddington, *The Expanding Universe* (Cambridge: Cambridge University Press, 1933).
2　For example, Harlow Shapley, in his book *Galaxies* (Oxford: Oxford University Press, 1973). This notion of the Milky Way as an unusually large spiral persisted into the 1990s in such recent reference works as *Encarta 94*.

Bibliography

These are the books that I read and used during the preparation of the book you now hold. The list is not a complete guide to everything written about the Universe, but if you wanted to dive in anywhere here and follow up the references, and the references contained in those books, and so on, you could probably spend the rest of your life happily probing mankind's understanding of the mysteries of nature. Some of the books are like old friends; others are useful because they provide one particular insight into one particular puzzle. Most are accessible to anyone who has read my own book, but a few, each marked with an asterisk (*), might be intimidating either on the grounds that they contain extensive mathematical equations or because they assume a detailed background knowledge of physics or astronomy. With that warning, feel free to read and enjoy. And remember:

> The true delight is in the finding out,
> rather than in the knowing
>
> Isaac Asimov

Abell, George, *Exploration of the Universe*, fourth edition (Philadelphia: Saunders, 1982).
A college text that covers everything from the Greeks to the Big Bang and sets in context just about all we know about the Universe as it is today. The book does not include anything on the inflationary cosmology, which is too new to have filtered into many college texts yet, but it is an excellent place to find out more about the Universe at large than I have had space to deal with here.

Atkins, Peter, *The Second Law* (New York: Scientific America/W. H. Freeman, 1984).
A well-illustrated and non-mathematical introduction to the supreme law of nature, the second law of thermodynamics.

Barrow, John and Tipler, Frank, *The Anthropic Cosmological Principle* (Oxford: Oxford University Press, 1986).
An enormous book (over 700 pages) in which the authors digress on just

about every subject relevant to the theme of mankind and the Universe, including a good, detailed section on its ultimate fate. Quite technical in parts, but an absorbing read in others. Good to dip into in the library!

Bernstein, Jeremy, *Three Degrees Above Zero: The Bell Labs in the Information Age* (New York: Scribner's, 1984).
This 'profile' of the Bell Labs is built around the story of Arno Penzias, Robert Wilson, and the discovery of the cosmic microwave background radiation. A lovely book, strong on people and places and invaluable background to the science I have tried to describe here.

Bonnor, William, *The Mystery of the Expanding Universe* (New York: Macmillan, 1964).
Bonnor, himself a cosmologist, wrote this book in the early 1960s, not long before the accidental discovery of the cosmic microwave radiation, now generally accepted as the 'echo of the Big Bang'. His very clear account of the different models of the universe allowed by General Relativity, and of rival ideas, sums up the state of cosmology in those days, when it was all very much still a mathematical game – just before the cosmologists were shocked, by the discovery of the cosmic background radiation, into the realization that the equations they scribbled so happily on their blackboards really did have a direct relevance to the origin and evolution of the real Universe in which we live. A lovely book – not so much dated as a period classic.

Chandrasekhar, Subrahmanyan, *Eddington* (Cambridge: Cambridge University Press, 1983).
The subtitle of the book – *The Most Distinguished Astrophysicist of His Time* – gives you a feel for the author's approach. Chandrasekhar studied under Eddington, and this little monograph is a tribute to the great man, in the form of lectures delivered to Trinity College, Cambridge, in 1982, the centenary of Eddington's birth. Full of anecdotal biography and a pleasure to read. Brief, but worth digging out simply for the description of the work leading up to the 1919 eclipse expedition that confirmed the prediction that light must bend as it passes near the Sun.

Davies, Paul, *Space and Time in the Modern Universe* (Cambridge: Cambridge University Press, 1977).
A delightful little book, which covers relativity and distorted spacetime, black holes, the traditional interpretation of thermodynamics and the relationship between life and the Universe, all in the space of 222 pages. Very well written; only slightly out of date.

* —, *The Forces of Nature* (Cambridge: Cambridge University Press, 1979).
A very clear account of the concepts underlying the modern understanding of the world within the atom – particles, fields and quantum theory. Light on math; strong on the 'feel' of the subject. My favourite Paul Davies book, and a good place to find out more about symmetry and unification of the forces.

—, *Superforce* (London: Heinemann, 1984).
Davies's best 'popular' book, covering much the same ground as *The Forces of Nature* but for an audience with less specialized knowledge of physics. Free from some of the constraints of writing textbook physics, Davies leaps off into more speculative realms in the later parts of the book and discusses subjects such as antigravity, the 'holistic' view of nature and even (in passing) astrology. If you read this and *The Forces of Nature* you'll get a good idea of what physicists are sure of, and also a glimpse of some of their wilder flights of fancy.

Eddington, Arthur, *The Nature of the Physical World* (Cambridge: Cambridge University Press, 1928).
Although written seventy years ago, Eddington's book is still worth reading for his discussion of the basic philosophy of the scientific approach and his still-clear explanation of concepts such as the thermodynamic arrow of time and the gravitational curvature of space.

Ferris, Timothy, *The Red Limit*, revised and updated edition (New York: Quill, 1983).
Subtitled *The Search for the Edge of the Universe*, this book provides a graphic overview of the historical development of the observations which have established the nature of the expanding Universe. It has very little about the origin of the elements or the fireball era, 'standard model' doesn't even appear in the index, and there is nothing on inflation, but Ferris provides a wealth of information and anecdote about the characters involved in probing out as far as our telescopes can see. Beautifully written, and a delight to read.

French, A. P. (editor), *Einstein: A Centenary Volume* (Cambridge, Mass.: Harvard University Press, 1979).
A collection published to mark the centenary of Einstein's birth, this volume includes reminiscences from his friends and colleagues, biography, outlines of Einstein's major contributions to science, and extracts from his own writings. Literally something for anybody with an interest in science, and a delight to dip into.

Gamow, George, *The Creation of the Universe* (New York: Viking, 1952).
A very readable, entertaining account of the early version of what became

the Big Bang model of cosmology. As well as providing a fascinating insight into Gamow's thinking, the book includes an equation for the temperature of the Universe at any epoch. Although Gamow didn't bother, in the book, to put the numbers into the equation, it does, in fact, 'predict' the existence of a cosmic background of radiation today with a temperature of a few K – a prediction published, not just in some obscure technical journal (though it was published in that form too) but in a popular book that sold in large quantities throughout the 1950s. And yet, when the predicted radiation was discovered in the 1960s, the pioneering prediction of Gamow and his colleagues had been forgotten by the experts.

—, *Mr Tompkins in Paperback* (Cambridge: Cambridge University Press, 1967).
Reprints of stories originally published in the 1940s, in which the mythical Mr Tompkins explores the worlds of relativity (including cosmology) and quantum physics. Updated by Gamow to take account of developments in the 1950s and early 1960s. Still great fun, and a painless way to absorb some of the key concepts of modern physics.

—, *My World Line* (New York: Viking, 1970).
Autobiography from 'Geo.', possibly the most versatile scientist of modern times, who participated actively in the three major scientific revolutions of the twentieth century – in quantum physics, cosmology and molecular biology. All this without being able to spell or, literally, to get his sums right. Very enjoyable stuff from a master storyteller.

*Gibbons, G. W., Hawking, S. W. and Siklos, S. T. C. (editors), *The Very Early Universe* (Cambridge: Cambridge University Press, 1983).
Definitely not for the mathematically fainthearted, this volume reports on a 'workshop' on inflation and associated ideas held at Cambridge University in 1982. A very important landmark in the history of the search for the Big Bang, and a good place for the mathematically inclined to get a feel of the excitement raised by the idea of inflation in the early 1980s. Contributions from Guth, Hawking, Linde, Rees and many others give you a ringside seat as new ideas unfold.

Gribbin, John, *In Search of Schrödinger's Cat* (New York/London: Bantam/Corgi, 1984).
The story of the development of quantum physics – the astonishingly successful, but non-commonsensical, theory of the word of the very small – during the twentieth century. Particle-wave duality, quantum uncertainty and the rest explained in (I hope!) readable fashion.

—, *Blinded by the Light* (New York: Harmony, and London: Bantam, 1991).
A more detailed account of the workings of the Sun (and other stars), including a discussion of the way WIMPs may solve a long-standing solar puzzle.

—, *In Search of the Edge of Time* (London: Bantam, and New York: Harmony, 1992).
General relativity pushed to its limits to explain the nature of black holes and the possibility of constructing time machines, all in line with the laws of physics as understood today.

Gribbin, John and Rees, Martin, *Cosmic Coincidences* (New York: Bantam, and London: Black Swan, 1990).
Subtitled 'the stuff of the Universe', this is our account of, among other things, the detailed nature of the candidates for the dark matter that keeps the Universe closed.

*Harrison, Edward, *Cosmology* (Cambridge: Cambridge University Press, 1981).
The author says his aim was to reach an audience 'at a level that is understandable to a college student who is not necessarily majoring in a natural science', and he mostly succeeds in this aim. A very good jumping-off point if you want to investigate cosmology properly, because he also includes copious references to the books and other publications containing the key ideas. But also a good book to dip into, with many juicy tidbits tucked away as 'reflections' at the end of each chapter. If you were to read just one book on cosmology (apart from my own!), then this is the one I would recommend.

Hoyle, Fred, *Galaxies, Nuclei and Quasars* (London: Heinemann, 1965).
Published just as the 3 K microwave background was being discovered, and mostly out of date, especially in terms of Hoyle's espousal of the Steady State theory. But worth seeking out for the last chapter alone, which summarizes the B^2FH scenario of the origin of the elements.

Hubble, Edwin, *The Realm of the Nebulae* (New York: Dover, 1958).
Originally published in 1936, Hubble's best-known book captures the freshness of the new discoveries about the nature of the Universe made in the 1920s and early 1930s. Very clear and well written, but remember that the distance scale of the Universe has been revised upwards by several notches since Hubble's day.

Kaufmann, William, *Universe* (New York: Freeman, 1985).
A nicely written and well-illustrated guide to astronomy, but one which, for my taste, puts too much emphasis on stars and planets and not enough on

the Universe at large. Designed for use in teaching a general science course, and worth dipping into if you can find it in a library. Shu's book is better.

Layzer, David, *Constructing the Universe* (New York: Scientific American Books, 1984).
A very good, beautifully illustrated account of cosmological ideas from Aristarchus to Einstein, marred by one flaw. Layzer is an advocate of a minority interpretation of the cosmological equations called the 'cold Big Bang' model. Instead of telling you this and offering the evidence for and against both his favoured model and the standard model, he presents the cold Big Bang as gospel and dismisses the standard model as old hat. This is a somewhat misleading representation of how cosmology is viewed by most people today. But if you treat this part of the book as a respectable minority view, and remember that the standard model is much stronger and more widely accepted than Layzer leads you to believe, the rest can be taken very much as the received wisdom today.

*Linde, Andrei, *Inflation and Quantum Cosmology* (London: Academic Press, 1990).
A 'horse's mouth' guide to inflation from one of the main proponents of the new ideas in cosmology. A specialist text that pulls no mathematical punches, but in which the equations are separated by very clear text and with the intriguing bonus of a set of Linde's own cartoon drawings illustrating key concepts.

*Narlikar, Jayant, *The Structure of the Universe* (Oxford: Oxford University Press, 1977).
A very thorough account of the present understanding of the Universe (but written before the idea of inflation surfaced) that includes much more about the variety of possible universes and the observations that may decide which kind we inhabit than I have had space for here. More technical or mathematical items are set off from the main text in boxes, which helps the readability, but this is still not a book for the casual reader.

—, *Introduction to Cosmology* (Boston: Jones and Bartlett, 1983).
The best no-punches-pulled textbook for anyone with a serious interest in cosmology and a thorough grounding in mathematics.

Novikov, I. D., *Evolution of the Universe* (Cambridge: Cambridge University Press, 1983).
Novikov, one of the senior Soviet cosmologists, came close to discovering the cosmic background radiation. His guide to cosmology is doubly interesting as a Russian view, which places the historical emphasis rather differently from

most Western accounts. Unfortunately, the translation is not everything it should be, and the result doesn't always read smoothly. But well worth investigating if you want to know more about the standard model. Fairly accessible, but perhaps meriting half an asterisk.

Overbye, Dennis, *Lonely Hearts of the Cosmos* (New York: HarperCollins, 1991).
A wonderful, great fat book about cosmologists in general, Allan Sandage in particular, and the search for the true value of Hubble's constant. Very much Overbye's personal view, idiosyncratic and biased towards the cosmologists he has met personally, but great entertainment.

Pais, Abraham, *Subtle is the Lord* (Oxford: Oxford University Press, 1982).
The definitive scientific biography of Einstein. Pais goes into a lot of scientific detail, but also provides a wealth of biographical information. He is very good at providing insight into *how* Einstein arrived at his great ideas, such as General Relativity, and I strongly recommend the book.

Pears, D. F. (editor), *The Nature of Metaphysics* (London: Macmillan, 1957).
A collection of essays based on talks for the Third Programme (as it then was) of the BBC, this relatively slim and well-written volume provides the best basic introduction to metaphysics that I know. Far more accessible and informative than many a weighty and learned academic tome.

Polkinghorne, J. C., *The Quantum World* (London: Longman, 1984).
A very neat little book – just a hundred pages, including a mathematical appendix and a glossary – that gets across most of the strangeness about quantum physics, and the philosophical discussions it has engendered. Especially interesting because the author was Professor of Mathematical Physics at the University of Cambridge from 1968 to 1979, then resigned from his post in order to train as a priest in the Church of England. He then became an assistant curate in Bristol. Recommended – not least to the army of physicists who use quantum physics today without ever pausing to wonder what it all really *means*.

Prigogine, Ilya and Stengers, Isabelle, *Order out of Chaos* (New York: Bantam, 1984).
Nobel laureate Prigogine made his name in the 1970s with new ideas about the meaning of the laws of thermodynamics and the nature of the Universe. This semi-popular account of his work provides the most complete and up-to-date explanation of thermodynamics for the non-specialist, and includes discussion of the arrow of time. A serious and deep book, which

rewards careful reading. For those in a hurry, though, Prigogine's ideas are summarized in snappy fashion in Alastair Rae's book, mentioned below.

Rae, Alastair, *Quantum Physics: Illusion or Reality?* (Cambridge: Cambridge University Press, 1986).
A readable little book on the puzzles of the physics of the world of the very small – atoms and particles. Mainly, as its title suggests, quantum physics; but including an excellent chapter on thermodynamics and the arrow of time, with the best concise summary I have seen of the latest ideas from Ilya Prigogine.

Rowan-Robinson, Michael, *The Cosmological Distance Ladder* (New York: Freeman, 1985).
The best and most accessible explanation of the way astronomers determine distances to remote objects, and thereby get clues to the age of the Universe and its ultimate fate. For a closed universe with omega equal to one, we need a rather lower value of the important Hubble parameter than Rowan-Robinson favours as the most likely value, but still within the 'error bars' of his calculation.

Sciama, Dennis, *Modern Cosmology*, revised edition (Cambridge: Cambridge University Press, 1973; reprinted in paperback, 1982).
A good place to look if you want to go a little deeper into the equations of cosmology than I have gone here, with some discussion of the different models allowed by the theories. The equations are not *too* intimidating, and there are plenty of helpful diagrams – but nothing, of course, on inflation, which hadn't been thought of in 1973. His older book *The Unity of the Universe*, published by Faber and Faber, is now, sadly, out of print, but worth looking out for in the libraries.

Shu, Frank, *The Physical Universe* (Mill Valley, California: University Science Books, 1982).
A great, big, delight of a book, aimed at liberal arts majors, which gives a solid overview of the Universe and our place in it for those who lack the mathematical background to appreciate, say, Narlikar's textbook.

*Silk, Joseph, *The Big Bang* (New York: Freeman, 1980).
A good introduction to the expanding Universe for anyone with a serious interest in the subject, including a great deal about how stars and galaxies form in the expanding Universe and a chapter on 'alternatives to the Big Bang' – the title is more than a little misleading and doesn't indicate the scope of the book. But you need some maths and physics to get the most out of it.

Weinberg, Steven, *The First Three Minutes* (London: Deutsch, 1977).
The best popular account of the standard model of the hot Big Bang, from a hundredth of a second after the moment of creation to the time when primordial nucleosynthesis came to an end and the hot fireball was left to expand into the Universe we know today.

Wright, Thomas, *An Original Theory of the Universe* (London: Macdonald, 1971).
A facsimile of the masterwork of the Durham philosopher, reprinted (with a new introduction by Michael Hoskin) in facsimile from the edition of 1750.

White, Michael and Gribbin, John, *Stephen Hawking* (London and New York: Viking, 1992).
Hawking's career and the significance of his own work, including his cosmological work, explained at what is (I hope) an even simpler level than that of the present book.

Index

Numbers in italics refer to captions.

PENGUIN ONLINE

READ MORE IN PENGUIN

In every corner of the world, on every subject under the sun, Penguin represents quality and variety – the very best in publishing today.

For complete information about books available from Penguin – including Puffins, Penguin Classics and Arkana – and how to order them, write to us at the appropriate address below. Please note that for copyright reasons the selection of books varies from country to country.

In the United Kingdom: Please write to *Dept. EP, Penguin Books Ltd, Bath Road, Harmondsworth, West Drayton, Middlesex UB7 0DA*

In the United States: Please write to *Consumer Sales, Penguin Putnam Inc., P.O. Box 12289 Dept. B, Newark, New Jersey 07101-5289.* VISA and MasterCard holders call 1-800-788-6262 to order Penguin titles

In Canada: Please write to *Penguin Books Canada Ltd, 10 Alcorn Avenue, Suite 300, Toronto, Ontario M4V 3B2*

In Australia: Please write to *Penguin Books Australia Ltd, P.O. Box 257, Ringwood, Victoria 3134*

In New Zealand: Please write to *Penguin Books (NZ) Ltd, Private Bag 102902, North Shore Mail Centre, Auckland 10*

In India: Please write to *Penguin Books India Pvt Ltd, 11 Community Centre, Panchsheel Park, New Delhi 110017*

In the Netherlands: Please write to *Penguin Books Netherlands bv, Postbus 3507, NL-1001 AH Amsterdam*

In Germany: Please write to *Penguin Books Deutschland GmbH, Metzlerstrasse 26, 60594 Frankfurt am Main*

In Spain: Please write to *Penguin Books S. A., Bravo Murillo 19, 1° B, 28015 Madrid*

In Italy: Please write to *Penguin Italia s.r.l., Via Benedetto Croce 2, 20094 Corsico, Milano*

In France: Please write to *Penguin France, Le Carré Wilson, 62 rue Benjamin Baillaud, 31500 Toulouse*

In Japan: Please write to *Penguin Books Japan Ltd, Kaneko Building, 2-3-25 Koraku, Bunkyo-Ku, Tokyo 112*

In South Africa: Please write to *Penguin Books South Africa (Pty) Ltd, Private Bag X14, Parkview, 2122 Johannesburg*

READ MORE IN PENGUIN

SCIENCE AND MATHEMATICS

The Character of Physical Law   Richard P. Feynman

'Richard Feynman had both genius and highly unconventional style . . .
His contributions touched almost every corner of the subject, and have had
a deep and abiding influence over the way that physicists think' – Paul
Davies

A Mathematician Reads the Newspapers   John Allen Paulos

In this book, John Allen Paulos continues his liberating campaign against
mathematical illiteracy. 'Mathematics is all around you. And it's a great
defence against the sharks, cowboys and liars who want your vote, your
money or your life' – Ian Stewart

Bully for Brontosaurus   Stephen Jay Gould

'He fossicks through history, here and there picking up a bone, an imprint,
a fossil dropping, and, from these, tries to reconstruct the past afresh in all
its messy ambiguity. It's the droppings that provide the freshness: he's as
likely to quote from Mark Twain or Joe DiMaggio as from Lamarck or
Lavoisier' – *Guardian*

Are We Alone?   Paul Davies

Since ancient times people have been fascinated by the idea of
extraterrestrial life; today we are searching systematically for it. Paul
Davies's striking new book examines the assumptions that go into this
search and draws out the startling implications for science, religion and our
world view, should we discover that we are not alone.

The Making of the Atomic Bomb   Richard Rhodes

'Rhodes handles his rich trove of material with the skill of a master
novelist . . . his portraits of the leading figures are three-dimensional and
penetrating . . . the sheer momentum of the narrative is breathtaking . . . a
book to read and to read again' – *Guardian*

READ MORE IN PENGUIN

SCIENCE AND MATHEMATICS

Bright Air, Brilliant Fire   Gerald Edelman

'A brilliant and captivating new vision of the mind' – Oliver Sacks. 'Every page of Edelman's huge wok of a book crackles with delicious ideas, mostly from the *nouvelle cuisine* of neuroscience, but spiced with a good deal of intellectual history, with side dishes on everything from schizophrenia to embryology' – *The Times*

Games of Life   Karl Sigmund
Explorations in Ecology, Evolution and Behaviour

'A beautifully written and, considering its relative brevity, amazingly comprehensive survey of past and current thinking in "mathematical" evolution . . . Just as games are supposed to be fun, so too is *Games of Life*' – *The Times Higher Education Supplement*

The Artful Universe   John D. Barrow

In this original and thought-provoking investigation John D. Barrow illustrates some unexpected links between art and science. 'Full of good things . . . In what is probably the most novel part of the book, Barrow analyses music from a mathematical perspective . . . an excellent writer' – *New Scientist*

The Doctrine of DNA   R. C. Lewontin

'He is the most brilliant scientist I know and his work embodies, as this book displays so well, the very best in genetics, combined with a powerful political and moral vision of how science, properly interpreted and used to empower all the people, might truly help us to be free' – Stephen Jay Gould

Artificial Life   Steven Levy

'Can an engineered creation be alive? This centuries-old question is the starting point for Steven Levy's lucid book . . . *Artificial Life* is not only exhilarating reading but an all-too-rare case of a scientific popularization that breaks important new ground' – *The New York Times Book Review*

READ MORE IN PENGUIN

SCIENCE AND MATHEMATICS

About Time   Paul Davies

'With his usual clarity and flair, Davies argues that time in the twentieth century is Einstein's time and sets out on a fascinating discussion of why Einstein's can't be the last word on the subject' – *Independent on Sunday*

Insanely Great   Steven Levy

It was Apple's co-founder Steve Jobs who referred to the Mac as 'insanely great'. He was absolutely right: the machine that revolutionized the world of personal computing was and is great – yet the machinations behind its inception were nothing short of insane. 'A delightful and timely book' – *The New York Times Book Review*

Wonderful Life   Stephen Jay Gould

'He weaves together three extraordinary themes – one palaeontological, one human, one theoretical and historical – as he discusses the discovery of the Burgess Shale, with its amazing, wonderfully preserved fossils – a time-capsule of the early Cambrian seas' – *Mail on Sunday*

The *New Scientist* Guide to Chaos   Edited by Nina Hall

In this collection of incisive reports, acknowledged experts such as Ian Stewart, Robert May and Benoit Mandelbrot draw on the latest research to explain the roots of chaos in modern mathematics and physics.

Innumeracy   John Allen Paulos

'An engaging compilation of anecdotes and observations about those circumstances in which a very simple piece of mathematical insight can save an awful lot of futility' – *The Times Educational Supplement*

Consciousness Explained   Daniel C. Dennett

'Extraordinary ... Dennett outlines an alternative view of consciousness drawn partly from the world of computers and partly from the findings of neuroscience. Our brains, he argues, are more like parallel processors than the serial processors that lie at the heart of most computers in use today ... Supremely engaging and witty' – *Independent*

BY THE SAME AUTHOR

In the Beginning   John Gribbin
The Birth of the Living Universe

Ripples in space collected by the COBE (Cosmic Background Explorer) satellite in 1992 clearly confirmed current ideas about the Big Bang. But why do matter and nature's fundamental forces seem specially designed to produce our kind of Universe? Some scientists see the hand of God; others call this a non-question; but Gribbin suggests a deeply satisfying new answer. Going far beyond the Gaia hypothesis that the Earth is a single living organism, he claims that galaxies may 'operate as supernova nurseries', that one universe can 'bud' from star-death and 'black hole bounce' into another, and that such 'offspring' are being steadily refined by evolution.

In Search of the Double Helix   John Gribbin
Quantum Physics and Life

Scientists now understand the fundamental secrets of life. Quantum effects cause tiny genetic mutations, transmitted by DNA, which fuel the struggle for survival among plants and animals. *In Search of the Double Helix* explains how such processes interlock. John Gribbin gives an account of the fierce (and sometimes unscrupulous) races to determine the structure of DNA and crack the ultimate code. Today, he argues, even analysis of amino acids in the blood confirms basic Darwinian principles and reveals how astonishingly close we are to gorillas and chimpanzees. His book offers an ideal overview.

BY THE SAME AUTHOR

In Search of the Edge of Time   John Gribbin
Black Holes, White Holes, Wormholes

The phenomena now known as black holes were described as early as 1783 and dismissed as the fruit of idle speculation – invisible stars sounded just too implausible to be taken seriously. It was only with the development of radio astronomy, relativity theory and mathematical models of warped spacetime that their true significance became clear. Today, writes John Gribbin, 'virtually all astrophysicists regard black holes as a natural feature of our Universe'. Many believe they can function as tunnels leading to other times and other places and that they contain the key to the Big Bang; Stephen Hawking sees them as 'wormholes' linking mother and baby universes. Details of such theories are set out in this enthralling book, a guided tour through a still emerging cosmos of neutron and X-ray stars, white dwarfs, quasars and pulsars.

'Fascinating ... Gribbin's thought-provoking book is written in the smooth, easy style of professional science journalism' – *The New York Times Book Review*

READ MORE IN PENGUIN

The Matter Myth Paul Davies and John Gribbin

Recent developments at the frontiers of science are challenging our views about ourselves and the nature of the cosmos as never before.

In this sweeping survey, acclaimed science writers Paul Davies and John Gribbin examine the revolutionary transformation that is currently overtaking scientific thinking. From the weird world of quantum physics and the theory of relativity to the latest ideas about the birth of the cosmos, they find evidence for a massive paradigm shift. Theories of black holes, cosmic strings, wormholes, solitons and chaos challenge common-sense concepts of space, time and matter and demand a radically new world-view. Here is a truly fascinating advance glimpse of twenty-first-century science.

The Stuff of the Universe John Gribbin and Martin Rees
Dark Matter, Mankind and Anthropic Cosmology

In trying to make sense of our relationship with the cosmos, scientists have concluded that most of the universe is made up of so-called 'dark matter', the controlling factor in its dynamics, structure and eventual fate. In this illuminating account leading science writer John Gribbin and eminent physicist Martin Rees give us the most comprehensive and accessible treatment yet of the major theories and the latest advances in under-standing the nature of dark matter, which lead on to the monumental question of why the universe is the way it is.

'The great question of Life, the Universe and Everything ... a pleasure to read' – Tim Radford in the *Guardian*

READ MORE IN PENGUIN

Richard Feynman: A Life in Science
John Gribbin and Mary Gribbin

'One of the most influential and best-loved physicists of his generation
... This biography is both compelling and highly readable' – Michael
White in the *Mail on Sunday*

Magnificently charismatic and fun-loving, Richard Feynman brought
a sense of adventure to the study of science. As well as leaving his
mark on nearly every aspect of modern physics, he was a hugely
popular and respected teacher. His extraordinary career included wartime work on the atomic bomb at Los Alamos and a profoundly
original theory of quantum mechanics, for which he won the Nobel
prize. In 1986 he came to widespread public attention during the
enquiry into the Challenger disaster when he proved conclusively that
its cause was due to the effect of cold on the shuttle's rubber sealings.

Skilfully interweaving personal anecdotes, writings and recollections
with narrative, John and Mary Gribbin reveal a man of startling
originality who had an immense passion for life.

Stephen Hawking   Michael White and John Gribbin

'Few scientists become legends in their own lifetime. Stephen
Hawking is one. It's good to have this well-documented and
immensely readable biography to remind us that the media-hyped
"mute genius in the wheelchair" is in fact a sensitive, humorous,
ambitious and occasionally wilful human being' – Paul Davies in *The
Times Higher Educational Supplement*